Nuclear Or Not?

Energy, Climate and the Environment Series
Series Editor: David Elliott, Emeritus Professor of Technology, Open University, UK

Titles include:

David Elliott (*editor*)
NUCLEAR OR NOT?
Does Nuclear Power Have a Place in a Sustainable Future?

David Elliott (*editor*)
SUSTAINABLE ENERGY
Opportunities and Limitations

Horace Herring and Steve Sorrell (*editors*)
ENERGY EFFICIENCY AND SUSTAINABLE CONSUMPTION
The Rebound Effect

Matti Kojo and Tapio Litmanen (*editors*)
THE RENEWAL OF NUCLEAR POWER IN FINLAND

Antonio Marquina (*editor*)
GLOBAL WARMING AND CLIMATE CHANGE
Prospects and Policies in Asia and Europe

Catherine Mitchell
THE POLITICAL ECONOMY OF SUSTAINABLE ENERGY

Ivan Scrase and Gordon MacKerron (*editors*)
ENERGY FOR THE FUTURE
A New Agenda

Gill Seyfang
SUSTAINABLE CONSUMPTION, COMMUNITY ACTION AND
THE NEW ECONOMICS
Seeds of Change

Joseph Szarka
WIND POWER IN EUROPE
Politics, Business and Society

Energy, Climate and the Environment
Series Standing Order ISBN 978–0–230–00800–7 (hb) 978–0–230–22150–5 (pb)

You can receive future titles in this series as they are published by placing a standing order. Please contact your bookseller or, in case of difficulty, write to us at the address below with your name and address, the title of the series and the ISBN quoted above.

Customer Services Department, Macmillan Distribution Ltd, Houndmills, Basingstoke, Hampshire RG21 6XS, England

Nuclear or Not?

Does Nuclear Power Have a Place in a Sustainable Energy Future?

Edited by

David Elliott
Emeritus Professor of Technology Policy, The Open University, UK

First published 2007
Published in paperback 2010 by
PALGRAVE MACMILLAN

Palgrave Macmillan in the UK is an imprint of Macmillan Publishers Limited,
registered in England, company number 785998, of Houndmills, Basingstoke,
Hampshire RG21 6XS.

Palgrave Macmillan in the US is a division of St Martin's Press LLC,
175 Fifth Avenue, New York, NY 10010.

Palgrave Macmillan is the global academic imprint of the above companies
and has companies and representatives throughout the world.

Palgrave® and Macmillan® are registered trademarks in the United States,
the United Kingdom, Europe and other countries

ISBN 978–0–230–50764–7   hardback
ISBN 978–0–230–24173–2   paperback

This book is printed on paper suitable for recycling and made from fully
managed and sustained forest sources. Logging, pulping and manufacturing
processes are expected to conform to the environmental regulations of the
country of origin.

... ... for this book is available from the British Library.

... ... y of Congress.

London Borough of Enfield	
91200000039833	
Askews	May-2010
333.7924	£19.99

Contents

List of Tables

List of Figures

Note on Units

Power units: The power-using or generating capacity of devices is measured in watts or W, or more usually kilowatts or kW (1 kW = 1000 W). Larger units are megawatts or MW (1000 kW), gigawatts or GW (1000 MW) and terawatts or TW (1000 GW).

Energy units: The kilowatt-hour (kWh) is the standard unit by which electricity is sold – 1 kWh is the energy produced/consumed when a 1 kW rated generator or energy-consuming device runs for one hour. A megawatt-hour (MWh) is 1000 kWh. Similarly 1000 MWh = 1 GWh and so on.

However, some analysts sometimes use the basic physical unit for 'work', the joule (J) or multiples of joules. One watt is one joule per second, so a kilowatt-hour is 3,600,000 joules, and the joule is thus a very small unit. Hence large multiples are common, e.g. peta joules or PJ (1000 tera joules) and exa joules or EJ (1000 peta joules).

Preface

Energy policy in Britain is undergoing a transition as profound as that which occurred a quarter of a century ago when North Sea oil and gas supplanted coal as primary sources of fuel and power. Now, as the North Sea supplies of oil and gas begin to decline, the country is facing a set of energy problems. There is the problem of shortage of electricity generating capacity as ageing coal and nuclear plants close. There is the issue of energy security that comes with increasing dependence on overseas supplies. Additionally there is the environmental problem created by the country's contribution to global warming (2% of the global total). Together these problems have ushered in a period where potential instability and uncertainty of energy costs and supply have placed energy policy high on the political agenda. From a period of relative quiet and consensus on the energy front in Britain there has emerged a transitional state of anxiety and conflict over energy security and its economic and environmental costs.

It is remarkable but rather little remarked that a major key to solving these problems would be a much more vigorous commitment to ensuring energy efficiency and conservation. Of course, energy efficiency is always an obligatory element in any energy strategy and considerable gains are being made with only a 2% increase in energy demand compared to a 21% rise in GDP over 1997–2005. A myriad ways of energy saving are explored in the 2006 Energy Review (Department of Trade and Industry, 2006). But, according to a House of Commons Environmental Audit Committee report, 'far greater political leadership is required and far higher priority to energy efficiency...' if reductions in energy demand are to be achieved (2006, p. 20).

The political focus has been on energy *supply* rather than *demand*. The key issue here is the need to ensure continuing and sufficient supplies of oil and gas especially in the short term in order to bridge the emerging 'energy gap'. Looking further ahead over the next ten years or so the issue is whether and to what extent alternative and indigenous sources of energy can supplant fossil fuels in order to achieve energy security and a low carbon economy. Provided sufficient political and financial commitment is made now there are a number of technologies within reach (integrated gasification combined cycle power plants, passive, safe nuclear reactors, carbon capture and sequestration, distributed energy systems). In the longer term, beyond twenty years, a range of possible technologies such as hydrogen fuelled transport, photo-voltaic cells on a large scale (already an established technology, for example, in Germany) or nuclear fast-breeder or fusion reactors are 'within sight but not yet within easy reach' (Fells et al., 2005, p. 28) and would require substantial investment in research and development.

Meanwhile, the immediate focus is on the relative merits of two available alternatives, renewable and nuclear energy. And the debate hinges around the issue of whether the country should embark on a new fleet of nuclear reactors to replace those that are being shut down. The issue has profound technical, ethical and political dimensions. Those who favour nuclear energy see it as a complementary element in the energy mix providing a low carbon secure source of supply. Opponents see nuclear energy as a dangerous technology in competition with the renewables sector that provides sustainable energy. Thus, the conflict over energy policy has become largely polarised around the single issue of whether nuclear energy has a role in the future energy supply of the United Kingdom.

After a period of relative quiescence, nuclear issues have become a focal point of debate over energy and the environment. Nuclear conflict subsided after the controversy over the THORP reprocessing plant and the battles over nuclear waste proposals culminating in the refusal of planning permission for the underground rock laboratory at Sellafield in 1997. Since then the nuclear industry has been warily negotiating with environmentalists to try to find solutions to the problems of waste. It seemed as if the nuclear issue, once so prominent, had slipped down the environmental agenda. This peaceful period and putative consensus has been interrupted, possibly disrupted, by what some have rather excitedly called a 'nuclear renaissance'. This is characterised by a resurgent and confident nuclear industry supported by industrial interests, by eminent scientists including the government's chief scientist and the Royal Society and by some powerful politicians. Although the nuclear industry still faces some major economic problems (for example, British Energy had to be bailed out by the taxpayer and cost estimates for nuclear decommissioning continue to grow, at the latest they are £70 billion), there is a new confidence around the industry supported by some favourable opinion polls. For years nuclear was overwhelmingly opposed, but within three years, 2001–2004, a MORI poll showed an increase from 19% to 30% of the population favouring a replacement nuclear programme with 34% against. A Eurobarometer survey in 2005 indicated 44% in the United Kingdom

(EU average 37%) favoured nuclear energy. Even more promising for the nuclear industry was a survey conducted by the Tyndall Centre for Climate Change Research in 2005 showing 61% support for continuing nuclear power provided this was coupled with development of renewables. The downside was the strong preference for solutions other than nuclear power, promoting renewables (78%) or changes in lifestyle and energy efficiency (76%).

What can account for this sudden change in the political environment surrounding nuclear energy? One obvious explanation, as suggested above, is that as the industry's confidence has grown, public concerns about nuclear safety have diminished. The big battles were over a decade ago and Chernobyl is now a distant memory. The industry has had time to regroup. But, the revival needed a reason and an opportunity. The reason was that nuclear was able to look more and more like the solution to our energy and environmental problems. As North Sea oil and gas decline so increasing dependence on Middle Eastern and Russian sources may threaten the security of supply. Nuclear risk seems very small when compared to the possibility of the lights going out. The opportunity was presented by a more sympathetic political environment and the forum provided by successive energy reviews in 2002 and 2006 and a White Paper on energy in 2003 (Cabinet Office, 2002; Department of Trade and Industry, 2003, 2006).

The nuclear case is framed on four fronts, in terms of safety, security, cost and conservation. At first sight, in view of the problem of routine and accidental emissions of radioactivity, safety seems an improbable claim. But, the industry can point to an excellent safety record (compared, for example, to the coal industry) and its routine emissions have been consistently reduced in response to tighter regulatory standards. While opposed to further development of nuclear the Sustainable Development Commission concedes that 'UK civil nuclear power stations have a very good safety record' but warns that the accidents, though rare, 'are also one of the main reasons for public concern' (2006, p. 14). Security of supply is seen as a major benefit from nuclear since it is a UK based source of energy. As for cost and taking into account construction, operating and decommissioning the industry argues vigorously that a fleet of stations using proven technology 'can offer electricity at predictable and stable costs for up to 60 years of operation' (World Nuclear Association, p. 21). Above all, in terms of environmental conservation, nuclear is being presented as the answer to climate change. If the United Kingdom is to meet its modest national target of carbon emissions 20% below 1990 levels by 2010, let alone the 60% by

2050 put forward in the government's 2003 White Paper on Energy, it will have to rely on low carbon emitting energy technologies. One of these is nuclear energy.

The rising fortunes of nuclear are reflected in a changing policy context. The 2002 Energy Review did not rule out nuclear but was not especially encouraging. 'If other low carbon options were to prove difficult to develop, then the case for nuclear power would be strengthened' (Cabinet Office, 2002, p. 123). However, the nuclear option should be kept open. The subsequent White Paper (DTI, 2003) was, if anything, rather more negative on the subject of nuclear's future. 'Although nuclear power produces no carbon dioxide, its current economics make new nuclear build an unattractive option and there are important issues of nuclear waste to be resolved. Against this background, we conclude it is right to concentrate our efforts on energy efficiency and renewables' (p. 122). Contrast this with the altogether more sanguine pronouncement in the energy review only three years later, 'Government believes that nuclear has a role to play in the future UK generating mix alongside other low carbon generating options' (DTI, 2006, p. 113). In the interim, the economics had apparently become favourable, the problems of waste were being resolved and confidence could be placed in the security and safety standards enforced by national and international regulatory regimes.

Although nuclear new build became the most prominent issue politically, it was only part of the emerging energy mix. Renewables combine three major advantages: local availability thereby avoiding imports; continuing availability so avoiding resource depletion; and generally low carbon output. Renewables had been the chief focus of the 2002 review with its suggested target of 20% of electricity supply, albeit at about 5% higher cost. The target was endorsed in the White Paper which also indicated that, to achieve a 60% reduction in carbon emissions by 2050, renewables would need to contribute 30–40% of electricity. Starting from a trivial 3% today these aspirations represent a fundamental transformation in electricity supply in the United Kingdom. Renewables covers a diversity of sources. In the immediate future wind will provide the biggest source of renewable energy mainly from onshore sources but with an increasing amount coming from offshore. Looking further ahead biomass, wave power, tidal schemes (including ambitious projects such as the Severn Barrage) and so on will make an increasing contribution. The target is achievable but 'it will require a far greater degree of commitment in terms of implementation than has hitherto been demonstrated' (House of Commons, 2006, p. 24).

Renewables face problems of their own. They are not necessarily cheap when compared to fossil fuel systems. They are supported by the Renewables Obligation requiring electricity suppliers to provide a certain percentage from renewable sources (potentially rising to 20% proposed in the 2006 Energy Review). Cost comparisons are extremely difficult and subject to considerable fluctuation. The best that can be said is that renewables' costs are likely to experience long-term decline as economies are achieved. Renewables, notably wind power, are intermittent sources of supply though the problem is often greatly exaggerated. As the proportion of renewable power supply grows it becomes necessary to provide stand-by generation. In principle, this is little different from the stand-by capacity needed to cover maintenance and other outages in fossil fuel and nuclear supply systems. Renewable forms of energy also attract opposition for the putative amenity damage and environmental impact they cause. This has slowed progress in developing wind farms in areas of landscape value.

The future energy mix will also see a shift in emphasis away from large-scale power plants and long-distance transmission (electricity and gas grid) towards 'distributed energy' systems. Broadly speaking, these cover a variety of technologies including: small-scale plant connected into a local distribution network; combined heat and power (CHP) systems; and microgeneration (small installations serving buildings or communities). According to the 2006 Energy Review, 'A "distributed" system could fundamentally change the way we meet our energy needs, contributing to emissions reduction, the reliability of our energy supplies and potentially to more competitive energy markets' (DTI, 2006, p. 62).

The future may also see some revival in traditional technologies. One possibility, rapidly moving from fantasy to reality, is the concept of 'clean coal'. Coal still contributes around a third of electricity generation (and during the winter of 2005/2006 this rose to around a half) though, along with nuclear, it is set to decline as ageing plants are retired over the next few years. But coal-fired power stations are a major source of carbon emissions. These emissions can be reduced through more efficient combustion technology, through burning coal with biomass and through carbon capture and sequestration whereby carbon is literally buried underground, most probably in depleted oil fields such as those in the North Sea. Clean coal technologies could reduce the carbon emissions by as much as 80 or 90%.

In the more distant future new energy technologies such as hydrogen production and storage could provide for a wide range of uses including transport but these are a long way off at present. For the foreseeable future, the next fifteen to twenty years or so, it is inevitable there will be increasing reliance on imported oil and gas (90% of the total) combined with development of indigenous supplies of renewables. There is the question of what part, if any, clean coal and nuclear energy will play in the mix.

It is the question of nuclear energy which continues to set the terms of the political debate about future energy supply. Yet, looked at in the wider context, nuclear's role would appear to be marginal at best. However, the case for or against nuclear does illuminate the various aspects of the energy question. Among these the most important seem to be carbon reduction, energy diversity and system flexibility.

The case for nuclear rests, partly, on its potential contribution to carbon emissions reduction. Nuclear power produces 4.4 tC/GWh compared to 97 tC/GWh for gas and 243 tC/GWh for coal. It should be noted that some critics suggest the figure for nuclear will rise as more energy is used to fabricate fuel from lower grade uranium ore. Moreover, nuclear only provides electricity, a secondary form of energy. Consequently its contribution to emissions reduction is in the power generation sector which accounted for around 43 MtC in 2000. Nuclear has little direct impact on emissions coming from the transport sector which contributes one-fifth of emissions or the industrial, domestic and commercial sectors which account for a little under half the total emissions. It is estimated that if existing nuclear stations were replaced, carbon emissions by 2030 would be around 8 MtC lower, the equivalent of the emissions from twenty two 500 MW gas-fired stations (DTI, 2006, p. 17). Set against predicted total emissions of 144–48 MtC in 2020 this is a modest though not insignificant saving. The issue is whether savings of this order could be achieved by greater energy efficiency or through deployment of more renewable technologies to replace nuclear.

Apart from its role in reducing emissions nuclear's perceived future role is 'to maintain the diversity of our energy mix' (DTI, 2006, p. 8). The assumption here is that nuclear will help to contribute to energy security by reducing dependence on imports as well as offering competition within the electricity market. Although uranium is an imported fuel it only accounts for about 11% of costs and sources of supply are expected to be relatively assured in the long term. Although there is greater optimism about the competitiveness of nuclear and its ability to solve decommissioning and radioactive waste problems, any future development will depend on private sector finance. While the government will assist by easing planning and regulatory hurdles its enthusiasm falls far short of providing a secure market framework to encourage the necessary long-term investment. There is only the vague promise that 'government will engage with industry and other experts to develop arrangements for managing the costs of decommissioning and long term waste management' (DTI, 2006, p. 125). Despite the enthusiasm expressed for nuclear power, encouragement is rhetorical rather than financial.

On the aspect of flexibility it must be said that nuclear energy is an inflexible method of producing electricity. Developing nuclear power takes time. Accelerated planning, pre-licensing regulatory assessment and the use of standardised reactor design will all help to reduce the very long lead times and 'appraisal optimism' associated with earlier nuclear programmes. However, even if a replacement programme were instituted today (2006), it is unlikely that any electricity could be delivered from new build before around 2020. Once commissioned nuclear power stations are likely to operate up to fifty years or more. By the time the new fleet is operating and certainly during the lifetimes of the power stations there are likely to be alternative, more flexible, and possibly cheaper, systems of supply available. As a critical report by the Sustainable Development Commission puts it, a 'single-minded focus on one large solution could lead to a significant decrease in both political and economic attention for the wide variety of smaller solutions that we will need over the long-term to move to a low carbon economy' (2006, p. 12).

A commitment to new build would also help to sustain the dominance of large-scale centralised systems of supply. It is likely that new stations would be built at or near existing sites thereby capitalising on the infrastructure, transmission links and public acceptability existing in those locations. This has both technical and social consequences. At the technical level a centralised system might diminish the network reinforcement needed to cope with renewables and distributed energy. At the social level a centralised system which includes nuclear energy tends to place control in remote, secretive and authoritarian institutional structures. By comparison distributed networks offer the prospect of more localised and potentially democratic structures. Although there is a tendency for renewables to be dominated by large power companies, in principle, they could be managed through local or co-operative ownership as is the case in other countries.

Nuclear energy is an ethical issue. Its association with nuclear weapons, proliferation, terrorism and accidents on the scale of Chernobyl arouses considerable anxiety. Beyond that is the intergenerational effect that issues from the creation of nuclear waste. Some of these wastes remain highly radioactive for unimaginable periods of hundreds of thousands of years and a burden of cost, effort and risk is passed down the generations. The problem of existing wastes must be dealt with – but that cannot justify the deliberate creation of new wastes from a new build programme. The Committee on Radioactive Waste Management (CoRWM) has proposed deep geological disposal of existing wastes as soon as practicable while recognising that interim storage will be necessary for at least two generations. But it has also made it clear that 'the political and ethical issues raised by the creation of more wastes are quite different from those relating to committed – and therefore unavoidable – wastes' CoRWM, 2006, p.13). Apart from anything else new wastes would extend the timescales for implementation over long and essentially unknowable future periods.

Nuclear energy certainly could provide part of the answer to solving the problems of reducing carbon emissions and dependence on imported energy supplies. But, its contribution needs to be set against the opportunity costs of a new build programme and the ethical considerations it would raise. In that context it must be said that there appear to be alternatives which are more flexible and ethically acceptable. A return to nuclear would represent a return to an older technology with its attendant dangers and emphasis on centralised supply of power. Moreover, it would provide the illusion of a solution to the problems of global warming and energy security which required no fundamental changes in production or consumption. It is this business-as-usual aspect of nuclear that is its most insidious characteristic. According to Fells et al. (2005), beyond 2025 diminishing returns will make it increasingly difficult to achieve reductions in carbon output on the scale of 60% by 2050. It 'will require huge additional investment and, taken with the inexorable rise in transport emissions – particularly air transport – the long-term future looks less optimistic. The chances of achieving the 60% figure must be very slender indeed' (p. 32). The danger is that by focusing on nuclear we refrain from recognising the scale of the challenge we face and shirk our responsibility for dealing with it.

Andrew Blowers

References

Cabinet Office (2002) *The Energy Review,* A Performance and Innovation Unit Report, February.
Committee on Radioactive Waste Management (2006) *Managing our Radioactive Waste Safely: CoRWM's Recommendations to Government,* July.
Department of Trade and Industry (2003) *Energy White Paper, Our Energy Future – Creating a Low Carbon Economy,* Cm 5761, February.

Department of Trade and Industry (2006) *The Energy Challenge,* Energy Review Report, HMSO, Cm 6887, July.

Fells, A., I. Fells, and J. Horlock (2005) 'Cutting greenhouse gas emissions – a pragmatic view', *The Chemical Engineer,* July, pp. 28–32.

House of Commons Environmental Audit Committee (2006) *Keeping the lights on: Nuclear, Renewables and Climate Change,* Sixth Report of Session 2005–06, Volume 1, April.

Sustainable Development Commission (2006) *The Role of Nuclear Power in a Low Carbon Economy,* London, March.

'The Energy Challenge', Energy Review Report, Department of Trade and Industry, London, July 2006.

World Nuclear Association (undated), *The New Economics of Nuclear Power,* WNA, London.

Series Editor Preface:
Energy, Climate and the Environment

Concerns about the potential environmental, social and economic impacts of climate change have led to a major international debate over what could and should be done to reduce emissions of greenhouse gases, which are claimed to be the main cause. There is still a scientific debate over the likely scale of climate change, and the complex interactions between human activities and climate systems, but, in the words of no less than Arnold Schwarzenegger, the Governor of California, *'I say the debate is over. We know the science, we see the threat, and time for action is now.'*

Whatever we now do, there will have to be a lot of social and economic adaptation to climate change – preparing for increased flooding and other climate related problems. However, the more fundamental response is to try to reduce or avoid the human activities that are seen as causing climate change. That means, primarily, trying to reduce or eliminate emission of greenhouse gasses from the combustion of fossil fuels in vehicles and power stations. Given that around 80% of the energy used in the world at present comes from these sources, this will be a major technological, economic and political undertaking. It will involve reducing demand for energy (via lifestyle choice changes), producing and using whatever energy we still need more efficiently (getting more from less), and supplying the reduced amount of energy from non-fossil sources (basically switching over to renewables and/or nuclear power).

Each of these options opens up a range of social, economic and environmental issues. Industrial society and modern consumer cultures have been based on the ever-expanding use of fossil fuels, so the changes required will inevitably be challenging. Perhaps equally inevitable are disagreements and conflicts over the merits and demerits of the various options in relation to strategies and policies for pursuing them. These conflicts and associated debates sometimes concern technical issues, but there are usually also underlying political and ideological commitments and agendas which shape, or at least colour, the ostensibly technical debates. In particular, at times, technical assertions can be used to buttress specific policy frameworks in ways which subsequently prove to be flawed.

The aim of this series is to provide texts which lay out the technical, environmental and political issues relating to the various proposed policies for responding to climate change. The focus is not primarily on the science of climate change, or on the technological detail, although there will be accounts of the state of the art, to aid assessment of the viability of the various options. However, the main focus is the policy conflicts over which strategy to pursue. The series adopts a critical approach and attempts to identify flaws in emerging policies, propositions and assertions. In particular, it seeks to illuminate counter-intuitive assessments, conclusions and new perspectives. The aim is not simply to map the debates, but to explore their structure, their underlying assumptions and their limitations. Texts are incisive and authoritative sources of critical analysis and commentary, indicating clearly the divergent views that have emerged and also identifying the shortcomings of these views. However, the books do not simply provide an overview, they also offer policy prescriptions.

The present volume presents ample evidence of divergent views and perspectives in relation to nuclear power. Some are based on differing interpretations of data, but some involve conflicting strategic preferences, often reflecting underlying ideological commitments and prescriptions. The strength of some protagonists' belief in the validity of their case often makes it hard to separate out 'facts' and 'values' in the nuclear debate. While not claiming to have achieved complete objectivity, this book tries to provide a snapshot of some of the key arguments in a way which allows the reader to assess their validity.

David Elliott

Acknowledgements

Many of the chapters in this book are based on papers presented at the Open University Conference 'Nuclear or Not?' held in March 2005, updated to take account of subsequent developments. The author of Chapters 1 and 2 acknowledges as a major source, the excellent undergraduate lectures given at King's College London in the early 1990s by Walt Patterson, now an Associate Fellow of the Royal Institute of International Affairs and a nuclear physicist by training. Chapter 6 is a revised version of a paper published in *Energy and Environment* (Vol. 17, No. 2, 2006, pp. 175–80). The author is grateful for permission to re-use it. Chapter 16 is based on an academic working paper which appeared on a Cambridge University research web site, a version of which was also published by *Prospect* magazine. The author is most grateful to Robin Grimes, Malcolm Grimston, Samantha King, Werner von Lensa, David Hamilton and Waclaw Gudowski and to an anonymous referee from the University of Cambridge, Electricity Policy Research Group, for insightful and helpful comments. He is also most grateful to Simon Smith for the suggestion that he follow work on *nuclear renaissance* with consideration of a *nuclear enlightenment*. Finally he is also grateful to SKB Sweden for making available the image used for Figure 16.1.

All errors and unattributed opinions expressed in this book are those of the authors alone and, as such, no responsibility for such matters rests with those that have kindly provided assistance.

The editor is grateful to Sally Boyle of the Open University Department of Design and Innovation, who kindly agreed to produce the graphics.

Notes on Contributors

Prof. Andrew Blowers, OBE, was until his recent retirement, Professor of Social Sciences (Planning) in the Department of Geography at the Open University. He is now a Visiting Research Professor at the OU and a member of the government's advisory Committee on Radioactive Waste Management (CoRWM).

Prof. David Elliott is Emeritus Professor of Technology Policy at the Open University and co-director of the OU Energy and Environment Research Unit. He trained initially as a nuclear physicist and worked at the UK Atomic Energy Authority, Harwell. At the OU he has developed courses and research programmes on technological innovation, focusing in particular on renewable energy technology policy.

Dr Jonathan Scurlock is a Visiting Research Fellow in the Energy and Environment Research Unit at the Open University. A specialist in environmental science and energy policy, in particular renewable energy and the global carbon cycle, he currently works for North East Community Forests on development of biomass energy, biofuels and community wind power.

Dr Horace Herring is a Visiting Research Fellow in the Energy and Environment Research Unit at the Open University, working on sustainable energy policy issues, energy economics and environmental history.

Paul Allen is Development Director at the Centre for Alternative Technology in Wales, where he has been involved with the design, manufacture and installation of a wide variety of renewable energy projects and with issues related to development of sustainable futures.

Prof. Gregg Butler is Professor of Science in Sustainable Development at the University of Manchester and a consultant with Integrated Decision Management Ltd. He was a Director of UK Nirex from 1990 to 1994 and Deputy Chief Executive of BNFL from 1993 to 1996.

Grace McGlynn works for Integrated Decision Management Ltd based in Preston. Both she and Prof. Butler worked for many years in the nuclear industry, between them covering all aspects of the nuclear fuel cycle and its interaction with the environment, policy and politics.

Stephen W. Kidd is Director of Strategy and Research at the World Nuclear Association based in London.

Dr Ian Fairlie is an independent consultant on radioactivity in the environment. He was formerly an occupational health advisor to the Trades Union Congress, and has worked on radiation protection issues for the Green Party Group in the European Parliament. He served on the Secretariat of the government appointed Committee Examining Radiation Risks of Internal Emitters (CERRIE).

Dr David Lowry is a Visiting Research Fellow in the OU Energy and Environment Research Unit and an independent environmental policy consultant working with elected politicians of different parties and as a freelance journalist for many publications and media outlets.

Dr Catherine Mitchell is a Principal Research Fellow at the Centre for Management Under Regulation at the Warwick Business School, University of Warwick. She was seconded as a team member to the Performance and Innovation Unit's Energy Review in the Cabinet Office during 2001.

Bridget Woodman is a UKERC Research Fellow at the Centre for Management Under Regulation at Warwick Business School working on sustainable energy policy and regulation issues.

Antony Froggatt is an independent researcher and freelance writer in the energy policy field.

Godfrey Boyle is Senior Lecturer in the Department of Design and Innovation at the Open University and Director of the OU Energy and Environment Research Unit. He has special interest in energy modelling and has a project on PV solar developments in Germany.

Dr William J. Nuttall is at the Judge Business School, Cambridge University, where he is Senior Lecturer in Technology Policy.

Introduction

Dave Elliott

With the likely social and economic costs of climate change being taken increasingly seriously, the nuclear lobby is arguing that it has at least part of the answer since, unlike coal or gas fired power plants, nuclear plants do not generate carbon dioxide gas, the main greenhouse gas responsible for climate change.

Nuclear power had fallen from favour due to high costs, and concerns about plant safety and radioactive waste disposal. More recently security issues have come to the fore. However, if we have to phase out our use of fossil fuels to reduce the impact on the climate system, then, say the nuclear proponents, what other energy source is there? The renewable energy lobby argues that it has the answer – there are many new 'green' energy sources which can provide all the energy we need without any emissions, and they are developing rapidly, especially in countries which have decided to back off from nuclear power. In addition, there is the energy efficiency option – we can avoid wasting so much energy, so that it becomes easier to meet our needs from renewables sources.

The debate between these various options sometimes focuses on the problems each perceives with what the other side are offering, for example in relation to direct costs. Energy saving is usually seen as the more economic option, at least initially, although once all the easy and cheap savings have been made, the costs of making more will rise. While the supply side options are more expensive, it may turn out that there may not be much in it in terms of initial capital costs – nuclear power is currently expensive, but so are some renewable energy technologies, although both could get cheaper.

There are also technical arguments. Nuclear plants only produce electricity, whereas that is only about 30% of what we need; renewables can provide heat and transport fuels, as well as electricity. However, the counter argument is that some renewable energy sources are intermittent and are not reliable as electricity sources, and growing biofuels for vehicles would take up a lot of land space. There is thus plenty of room for debate.

The comparison gets more favourable to renewables when we look at some of the other major differences between these two types of energy supply technology. Nuclear plants need uranium to fuel them and inevitably produce dangerous long-lived wastes, whereas most renewables need no 'fuel' and produce no wastes, and so there are no fuel or 'backend' costs. Moreover, when nuclear plants come to the end of their working life, they must be decommissioned, which is a very expensive process – which generates yet more wastes. Nuclear facilities are also potential terrorist targets, whereas most renewables are unlikely to attract the attention of terrorists.

It is perhaps not surprising then that most people have opposed nuclear power – opinion polls over the years have typically indicated that 70–80% of those asked were against an expanded nuclear programme, while about the same number were in favour of renewables like wind power.

However, things may be changing. In the UK, opposition to a 'replacement' nuclear programme (replacing the UK's old plants as they are retired) has fallen in recent years, no doubt reflecting growing concerns about climate change. A Mori poll in 2001 found that 19% favoured a replacement programme, while 57% opposed it. By 2004 the figures had changed to 30% for and 34% against. The renewables lobby replies, if climate change is so important then that is all the more reason to expand support for renewables, since they are the best bet for a long-term sustainable energy future. Going back to nuclear power would simply divert resources from developing renewables. Moreover, why try to solve one environmental problem (climate change) by creating another (radioactive pollution), especially since long term we will have to switch over to the renewables, since there are only limited amounts of uranium available in the world?

So the debate goes on. On one side we have renewable energy protagonists arguing that renewables could supply 50% of total world energy requirements by 2050, compared to the 7% currently supplied by nuclear power. On the other, some nuclear proponents say that new nuclear technologies will be cheaper, cleaner and safer, and that in any case, the costs and risks of climate change outweigh the problems of nuclear power. Then again some say – why not have both? But, so the counter argument goes, we do not have the resources to do this – and there would be a risk of doing neither well.

Although there are many disputes along these and other lines, a rough consensus does seem to have emerged, at least in the UK, on the basic boundary conditions, with, for the purposes of our discussion, two polar options being seen as credible. They were well set out by the Royal Commission on Environmental Pollution in 2000 in its study

Energy – the Changing Climate, and they have been developed further in the Department of Trade and Industry's *Options for a Low Carbon Future* (DTI Economics Paper No 4 2003) and the linked Future Energy Systems papers *Options for a Low Carbon Future Phase 1 and Phase 2*. Some of the results are described in Boyles' contribution (Chapter 13) to this book.

To summarise, on one hand, it is fairly widely accepted that it should be technically possible for renewable energy technologies, coupled with energy efficiency, and the use of carbon capture and storage ('CCS'), to supply sufficient power to meet UK energy demand while reducing carbon emissions by 60% by 2050. This is despite the fact that, as the Sustainable Development Commission has noted, on current plans, the nuclear contribution will decline to under 3% by around 2024, and zero by around 2036, when the last remaining plant, Sizewell B, is expected to close. The use of geological sequestration (storage) of carbon dioxide in depleted oil and gas wells, and perhaps saline aquifers, would allow fossil fuels to supply about 50% of UK electricity without adding significantly to emissions, while renewables would supply most of the remainder, again without adding to emissions. Low carbon energy from Combined Heat and Power ('CHP') plants would make up the rest. No one says it will be easy – it would require a rapid ramping up of renewables and a commitment to efficiency, CHP and clean coal/CCS. But it is a credible option

On the other hand, it is also argued that the same sort of outcome could be attained by expanding nuclear power up to around a 40% contribution, along perhaps with a smaller contribution from renewables, plus an energy efficiency programme, but with no, or little, CCS. In between these extremes there are a range of mixes – for example with less nuclear but more CCS, or with more reliance conservation/efficiency. Essentially, the debate is, or at least ought to be, on which mix to choose, and on the practical and strategic viabilities of specific mixes.

This book will attempt to explore this debate. However, it is not meant as a technical treatise. Rather it aims to relay and review the views of some of the key protagonists. It will look at the views of those who are keen to promote nuclear expansion, at the challenges that face them, and at the views of those who wish to promote alternative approaches, and their limitations. It is based in part on papers presented at a one day conference in March 2005, organised by the Open University Energy and Environment Research Unit and the OU Department of Geography, under the title 'Nuclear or Not?' In addition some other papers were commissioned, mostly from people who attended the conference.

Although the OU conference looked critically at the issues surrounding the proposed expansion of nuclear power in the UK, it was not meant to be a rehearsal of pro- and anti-nuclear arguments, but rather it was an attempt to set the debate in the wider context of what is the best way to deal with energy supply and demand in relation to climate change. This book takes a similar approach.

The Nuclear debate

In 1998 the nuclear industry journal *Nucleonics Week* (22, October 1998) said, perhaps rather tongue in cheek, that 'nuclear needs climate change more than climate change needs nuclear' and that issue remains central. The nuclear protagonist claim that it can help respond to climate change since nuclear plants do not generate carbon dioxide. The opponents however raise a whole series of objections, most of them familiar from the nuclear debate over the last few decades. But the nuclear proponents argue that the threat of climate change may put some of these objections into a new perspective – and requires a new more objective analysis of the pros and cons of nuclear power.

In the first three papers in this book, to set the scene, Scurlock and Herring, provide an historical overview of nuclear technology and reactions to it. Moving up to date to the current debate, Butler and McGlynn (Chapter 4) and Allen (Chapter 5) then lay out views from, as it were, the two camps, as to what the issues are, and how they should be assessed. They also ask if there can be common ground – perhaps leading to a consensus approach. It is perhaps unsurprising that there remains a gap in the 'bridge building' process between these two positions – although, the nuclear objectors seem to be moving the debate on to the relative merits of nuclear and renewables, and the nuclear lobby now seem to be less prone to attack renewables as irrelevant. Indeed, as is illustrated by Kidd, who lays out a case for a new approach to nuclear power (Chapter 6), the nuclear lobby seems increasingly keen to see renewable as an ally. However, as my own paper illustrates (Chapter 7), this offer has not so far been reciprocated. Most renewable energy supporters fear that a return to nuclear will undermine the development of what they see as a truly sustainable energy future.

That is not to say that the 'traditional' objections to nuclear power-concerning safety, security and cost – do not still have some force. In the next set of papers (Chapters 8–11), Fairlie, Lowry, and Mitchell and Woodman, provide some examples of how some of these arguments have developed. In effect these are some of the key practical challenges that nuclear proponents have to face.

In the next selection of papers we move back to the strategic level, by looking at some national examples. Froggatt (Chapter 12) provides an overview of the situation in Europe, while Boyle (Chapter 13) compares the situation in Germany and the UK. France and Finland may currently be the exceptions in Western Europe, in pressing ahead with new nuclear plants, but some of the ex-Soviet states are still pro-nuclear. In addition the US government is seeking to revive its nuclear programme, and of course nuclear power is back on the agenda in the UK. In parallel, China, India and some other Asian countries are looking to nuclear expansion as one option for the future. Indeed, in Chapter 14, Kidd argues that it may be that nuclear powers' future lies in developments like this.

The final section tries to draw the review of issues and processes together, looking at the process and framing of the debate, the wider strategic issues, and prognoses for the future. In Chapter 15, Scurlock asks can nuclear power ever be seen as 'green' and concludes that it is unlikely. By contrast, in Chapter 16, Nuttall argues that a new awareness may emerge and indeed will be necessary if nuclear power is to expand. Clearly there are disagreements, with, at base, as I argue in the final Chapter, the issues being as much ideological as technical.

Nuclear proponents sometimes argue that it is 'ideological' to oppose nuclear, as they say some do, on the grounds that it is a key part of a centralised capitalistic consumer society. However you could argue that it is equally ideological to support nuclear power on these – or any other – social and political grounds. Nuclear technology is clearly not the only thing shaping modern industrial society, but for some it has come to symbolise many of the faults of that society and of the way it is developing. In Chapter 6, Kidd says 'To rule out any option through ideology is not appropriate', but to a degree, that is what we look to politicians, and the wider political system, to do. It is not just a simple technocratic issue: technology shapes society and society must try to shape technology.

Certainly we will have to move away from the simple assumption that we can have 'more of everything' in energy option terms, backing all the horses: we have to make choices. Hopefully the discussions in this book will be part of a process which will help to ensure that these choices are made in a more enlightened way.

This book attempts to bring together a range of views, with contributions from authors with sometimes strong views on either side of the nuclear debate. As editor, I have tried to ensure that a reasonable balance of opinions is covered, but inevitably this will not please everyone. For example, some might have welcomed more coverage of the pro-nuclear viewpoint, while others would expect something from anti-nuclear environmental organisations like Friends of the Earth and Greenpeace. However, rather than simply trying to 'map' the various lobbies, I have allowed a range of practitioners, academics and policy specialists to put their views, with the emphasis being on critical approaches, as opposed to simple reiterations of positions.

The preparation of the first (hardback) edition of this book in 2007 coincided with the first phase of the Energy Review carried out in 2006 by the UK government, with nuclear power being a major issue. Some authors were able to take account of some of the submissions to the consultation phase of that review, and the Preface and final section included comments on some of its conclusions.

The 2006 Energy Review led to a White Paper on Energy, published in 2007, which in effect reversed the conclusions of the 2003 White Paper on Energy, which had said that the current economics of nuclear power '*make it an unattractive option and there are also important issues of nuclear waste to be resolved*'. The 2007 White Paper concluded that '*new nuclear power stations would make a significant contribution to meeting our energy policy goals*'.

The new view might be seen as reflecting increased concerns about security of energy supplies (notably gas) and also increased awareness of the threat posed by climate change. The UK commitment to a 60% reduction in greenhouse gas emissions by 2050 was beginning to be seen as insufficient, with talk of an 80% cut perhaps being needed. The 2007 White Paper on energy was followed, in 2008, by a White Paper on nuclear power, which took this argument on board, and claimed that nuclear could deliver at least part of the cut in a cost effective way: 'An 80% reduction scenario using central fossil fuel assumptions with the option of new nuclear power stations would reduce GDP in 2050 by 1.6%. Placing a constraint on new nuclear power stations would increase the costs to 1.7% of GDP in 2050.'

In parallel, the government had to respond to a new European Commission directive on renewables, which called for the EU to expand renewables so that they supplied 20% of total EU energy by 2020, with each EU country being expected to make a contribution, in line with the existing levels they had so far achieved and potentials for expansion. The UK, which at that point was only getting around 2% of its energy from renewables, negotiated a lower target of 15%. Even so, that would represent a major expansion, and mainly in the electricity sector, since expansion in the heat and transport sectors was seen as much harder to achieve: perhaps 30–40% of UK electricity would have to come from renewable sources by 2020.

So once again we had nuclear and renewables lining up as candidates for providing carbon-free, or at least low carbon, energy. The government however was adamant that they were not rivals – we could and should have both. Not everyone in the renewables community or the wider environmental community was convinced, and the debate has continued over how best to secure a sustainable energy future, and the role that nuclear power might play within it. With, as this new edition goes to press, plans now emerging from private sector companies for nuclear projects in the UK, and sites being selected, the debate on the wisdom of this policy and on the viability of other approaches seems to be deepening and becoming more urgent. Hopefully this book will feed into the continuing debate.

Part I Setting the Scene

1

Nuclear Energy: An Introductory Primer

Jonathan Scurlock

Introduction

The process of nuclear fission ('splitting the atom' or, more precisely, 'splitting the atomic nucleus') releases immense amounts of energy. Under controlled conditions within a nuclear reactor, this process can release one million times more energy per atom than any chemical reaction, including combustion. Furthermore, this occurs without many of the pollutants associated with combustion, e.g. oxides of nitrogen, sulphur and carbon. So it is hardly surprising that over the past 60 years considerable efforts have been made to harness this theoretically efficient use of the Earth's energy resources.

This chapter provides a succinct overview of the science and technology underpinning nuclear power, aimed at non-specialist readers with a grasp of basic physics, in order to help them compare nuclear with other energy policy options. Here, and in the following short history of the industry, the environmental implications of nuclear power are mentioned only in passing, since they are covered extensively elsewhere in this volume. However, answers (or pointers to answers) may be found to some of the questions commonly asked, for example, about reactor types, economics, military–civil links, the rationale for reprocessing and so on.

Some nuclear physics

The *atom* was originally defined by the Greek philosopher Democritus as the smallest indivisible unit of matter, i.e. the smallest part into which an element can be divided without changing its nature. The atom itself is depicted as a number of shells of negatively charged *electrons* orbiting a *nucleus* – a cluster of positively charged *protons* and uncharged *neutrons*.

Although the protons repel one another due to their similar electrical charge, the more powerful but short-distance 'strong nuclear force' which acts between all neutrons and protons holds the nucleus together (Figure 1.1).

Consulting the periodic table of the elements shows that there are 92 different elements found occurring naturally, ranging from those with the lightest nuclei (hydrogen – atomic number 1, helium – atomic number 2, etc.) to those with the heaviest nuclei (protactinium – atomic number 91, uranium – atomic number 92). The *atomic number* corresponds to the number of protons in the nucleus, balanced in charge by an equal number of orbital electrons, which determine the physical and chemical properties of each element. Together with its 92 protons, the nucleus of the heaviest naturally occurring element, uranium (symbol U), contains 146 neutrons. The *atomic weight* of each element is the sum of its number of protons and neutrons, since the orbiting electrons have negligible mass: in the case of uranium, this is 92 + 146, i.e. 238. Uranium with this atomic weight is known as uranium-238, or ^{238}U.

The stability of the nucleus is governed by the balance of attractive and repulsive forces between the protons and neutrons, but the neutron to proton (N:P) ratio can vary only within certain limits. In the case of lighter nuclei such as carbon or oxygen, the N:P ratio is about 1.0, whereas for the heaviest elements the N:P ratio rises to about 1.5.

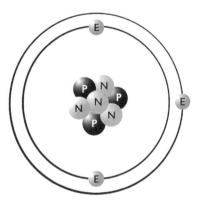

Figure 1.1   Generalised diagram showing component parts of the atom. For clarity, a relatively light nucleus is depicted (lithium, with atomic number 3 and atomic weight 7). N = neutron; P = proton; E = electron. Heavier radioactive nuclei such as uranium-235 contain many more protons and neutrons.

Since it is so massive, with a large number of protons and neutrons clumped together, the uranium nucleus is particularly sensitive to changes in the balance of these attractive and repulsive forces. Picture for yourself a seething mass of sub-atomic particles surging around in a state of constant tension! The physics of the uranium nucleus is further complicated by the existence of two naturally occuring *isotopes*, with the same atomic number but slightly different atomic weights. Although the vast majority of uranium nuclei are ^{238}U, 0.7% comprise an isotope known as uranium-235. ^{235}U has the same number of protons (92), but only 143 neutrons instead of 146. This small decrease in the N:P ratio makes ^{235}U even less stable than ^{238}U.

Coming apart at the seams

Unstable nuclei tend to emit particles or energy (*radioactive decay*), or more drastically, to break apart (*nuclear fission*) in an attempt to restore the balance of neutrons to protons. Radioactive decay may take the following forms:

- Alpha (α) – ejection of a helium nucleus, i.e. 2 protons and 2 neutrons. The relatively heavy α-particle can be stopped by a sheet of paper or the epidermis of the human skin, so it has only a short range, but a high probability of doing damage when it impacts on human cells. α-emitting nuclei therefore do not pose much danger outside the human body, but they can be very dangerous if inhaled or otherwise incorporated into tissue.
- Beta (β) – emission of an electron as a neutron changes into a proton, or emission of a *positron* (a light particle equivalent to an electron, but carrying a positive charge) as a proton changes into a neutron. β-particles (β^- and β^+) are small and light, with a low probability of collision with other nuclei as they pass through matter. β-radiation therefore has a longer range than α-radiation, but it can be stopped by a thin layer of metal foil.
- Gamma (γ) – high-energy photons emitted to get rid of excess energy in the nucleus. More energetic and potentially damaging than X-rays, γ-radiation has a long range and is only effectively stopped by massive shielding, e.g. a two-metre thickness of concrete.

All three kinds of radioactive decay may cause biological damage and require precautions to be taken with naturally occurring radioactive substances such as uranium and its compounds, as well as with any radioactive by-products of unstable nuclei. High doses of 'nuclear radiation' can kill a living cell, but much lower levels can damage genetic material and affect cell division (possibly leading to cancers). Details of radioactive decay chains, isotopic half-lives and biological effects of nuclear radiation can be found in many textbooks, but a good summary which remains a 'classic' of independent criticism is given by the UK Royal Commission on Environmental Pollution (RCEP, 1976).

More severe transformation of unstable nuclei may take place by the process of nuclear fission, whereby the nucleus divides into two substantial parts (rarely a 1:1 ratio; a 3:2 ratio being more likely). The *fission products* tend to be unstable also, with an excess of neutrons over protons, and so they are usually β- and γ-emitters. Strontium-90 (^{90}Sr) and caesium-137 (^{137}Cs) are typical products of uranium fission, with relatively long half-lives of 28 and 30 years, respectively.

'Spare' neutrons may also result from the fission process. Neutron 'radiation' can be quite penetrating, since these electrically neutral particles are slowed down only when they collide with other nuclei. Light nuclei such as hydrogen, carbon or oxygen are particularly effective at slowing down stray neutrons (the principle of a *moderator* – see below), which makes them easier for other nuclei to absorb. But since these light nuclei occur commonly in living tissue, it is also most important to shield the human body from neutrons.

Following the fission of a single nucleus of the unstable isotope uranium-235, two to three (on average about 2.5) stray neutrons are released, which may be 'captured' by other nuclei of ^{235}U. These, in turn, immediately become highly unstable, resulting in further *induced fissions*, with the accompanying release of further neutrons and energy as heat. Thus an initial spontaneous nuclear fission may result in 2 further induced fissions, then 4, 8, 16, 32, etc., resulting in a *chain reaction* (Figure 1.2). After 80 or so generations, the chain reaction produces a catastrophic release of energy – a nuclear explosion. The principle of the atom bomb was therefore how to set off a chain reaction to order (and not before!) – a function of the concentration of neutrons being produced, the proportion that result in induced fissions and their rate of loss from the outer surface of the mass of fissile material. In a nuclear weapon, this may be achieved by rapidly bringing together two smaller blocks of fissile material into a larger *critical mass*, within which the chain reaction will be sustained. However, civilian applications of nuclear energy require a more controlled chain reaction, maintained as a source of heat – without the possibility of an explosion.

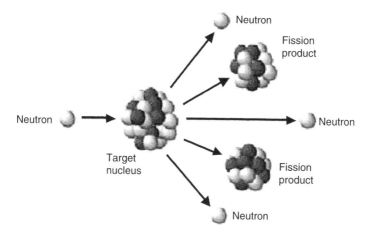

Figure 1.2   Neutron capture and induced fission, leading to a chain reaction with a *multiplication factor* of 2. Each fission yields two fission products and two more neutrons, which then induce two further fissions, etc.

A very complicated kettle

Within the core of most types of nuclear power reactors, rods of nuclear fuel containing ^{235}U are surrounded by a *moderator* material (such as graphite or water) which slows down stray neutrons, increasing the probability of further induced fissions. *Control rods* containing a neutron-absorbing material (commonly the boron found within boron steel) are also necessary to slow the chain reaction down to within manageable limits, or to shut the reactor down altogether in an emergency or for routine maintenance. A *coolant* is required to transfer the useful heat out of the reactor core: this may take the form of a gas such as carbon dioxide or helium, a liquid such as water or *heavy water* (acting as a dual-purpose coolant and moderator) or even a molten metal such as sodium. The hot coolant is then used to generate steam, which drives turbines as in any fossil fuel thermal power station (Figure 1.3).

A nuclear power station may therefore be thought of as a very complicated kettle coupled to conventional power station technology. However, the need to minimise handling of the nuclear fuel means that the reactor is loaded with fuel at the start of its life, and is then kept running as long as possible. Eventually the accumulation of fission products begins to interfere with reactor operation by absorption of neutrons and

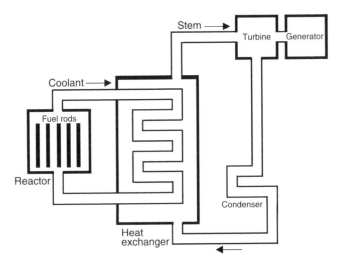

Figure 1.3   Simplified schematic diagram of a nuclear power station

can result in uncontrollable side reactions. At this stage, the fuel needs to be changed even though, typically, only about 1–2% of the ^{235}U has actually been consumed. Some reactors can be refuelled while running at reduced power, but other types need to be shut down completely for up to 2 months. The spent fuel is intensely radioactive, and ways have to be found to ensure its safe handling and ultimate disposal.

The need for enrichment

As explained above, natural uranium comprises two isotopes in the ratio 993 parts ^{238}U to 7 parts ^{235}U. In this form, natural uranium will not support a chain reaction unless a special neutron moderator is used. Heavy water (D_2O – water containing the heavy *deuterium* isotope of hydrogen) is suitable for this purpose, but it is complex and expensive to separate from ordinary water since only 0.02% of water molecules contains deuterium. In practice, in Canada and elsewhere, most heavy water reactors have shown poor economic performance, attributable in part to errors in the original estimates of heavy water separation costs.

The more commonplace alternative is to *enrich* the uranium by increasing the proportion of the more fissile isotope ^{235}U from 0.7% to 3–4%. This level is sufficient for most forms of nuclear fuel, but enrichment as far as even 93% is possible – as used in the first atomic bomb at Hiroshima.

While ^{235}U nuclei split after capture of a stray neutron, neutron capture by ^{238}U results in the rapid emission of 2 β^--particles, yielding a new type of artificial nucleus with atomic number 94 – plutonium (symbol Pu). ^{239}Pu sustains a chain reaction even better than ^{235}U, and is therefore an ideal material for nuclear weapons. However, it may also be used as a fuel in power reactors, as discussed below.

Uranium supply

The element uranium is found in various ores, which may be relatively rich in the metal (such as pitchblende) or much poorer and therefore only worth extracting when demand greatly exceeds supply. Mining, processing of the ore and enrichment are often referred to as the *front end* of the nuclear fuel cycle.

After mining the ore, the uranium is dissolved out, leaving a finely divided sand known as *uranium tailings*. This material has about the same bulk as the mined ore, and it still contains trace amounts of uranium together with all its natural decay products, such as radon-222 (which can be inhaled together with dust containing its own decay products) and radium-226 (a dangerous α-emitter, which may be incorporated into bone tissue by substitution for calcium).

Regrettably, hundreds of millions of tonnes of uranium tailings are piled up near mines in the western USA, Canada, Namibia, Eastern Europe and elsewhere, much of it exposed to wind and rain and therefore prone to unregulated dispersal. Some critics consider that uranium tailings represent the single largest waste disposal problem of the nuclear industry worldwide, ahead of concerns about spent fuel disposal. In one celebrated case, tailings from a uranium mill at Grand Junction, Colorado, were negligently allowed to be used for domestic construction work during the 1950s. The mistake was not noticed until 1966, and following legal action, the wastes were eventually removed and replaced from over 600 buildings between 1974 and 1988.

Mention should also be made of thorium, a rival fuel, which, although more abundant than uranium, is not so far used since it is not directly fissile, but which may be used, for example in fast neutron reactors, as uranium reserves are depleted.

Enrichment technology

Enrichment of uranium is necessarily based upon the physical properties of the ^{235}U and ^{238}U isotopes, since they are chemically identical.

The *calutron* was an early technology based upon firing a beam of uranium vapour through electric and magnetic fields. The deflection of the beam differed between the lighter and the heavier isotopes, providing a degree of separation. Although suitable for weapons, this proved to be a highly energy-intensive (and therefore pointless) way of making enriched uranium reactor fuel.

Gaseous diffusion methods developed more slowly under the Manhattan Project (see below). These utilise the tendency for one isotope of uranium vapour to migrate faster than the other through a porous nickel-based membrane under vacuum (McKay, 1984). However, since the degree of separation provided by a single diffusion step is tiny (1.004:1), a very long cascade (thousands of steps) of diffusion modules is required. As well as requiring a substantial energy input (up to 2000 MW for the former K-25 plant at Oak Ridge, Tennessee), gaseous diffusion plants were among the largest 20th-century industrial installations on Earth – many visible to spy satellites in space.

High-speed centrifugation of uranium hexafluoride gas is the latest enrichment technology giving the best *energy balance* (energy outputs:energy inputs). Like gaseous diffusion, this involves a cascade of many steps to gradually concentrate ^{235}U, but the technology is more compact (and thus more prone to unregulated proliferation to so-called unfriendly governments, terrorists, etc.).

Laser isotope separation using an electric field to discriminate between laser-excited uranium isotopes has been demonstrated only at laboratory level so far. It offers the potential of a cheap and energy-efficient single-stage enrichment process, but threatens a truly dangerous level of proliferation.

Very hot ashes

The *back end* of the fuel cycle entails handling, storing and/or reprocessing of the spent reactor fuel. Upon removal from the reactor, the fuel elements are still very hot (in the thermal sense) and require constant cooling. They are also intensely radioactive due to the accumulation of various fission products – in the absence of shielding, a fuel element at a distance of 10 metres would give you a lethal dose of radiation in about 20 minutes. Most of the fission products are economically useless, with relatively long half-lives, and must therefore be contained for long-term storage. However, there may be a case for separating the unused uranium (about 97% of the total) together with the plutonium (about 1%) for use as a fuel in fast breeder reactors, or for conversion to mixed-oxide

(MOX) fuel which can be used in conventional fission reactors. This is also how plutonium can be extracted for weapons. The argument over whether it is economically, operationally and environmentally desirable to *reprocess* spent nuclear fuel and *close the fuel cycle* continues to this day, with some national governments and industries carrying out reprocessing while others opt for long-term *dry storage* of the spent fuel.

Decommissioning

However, to debate optimisation of the fuel cycle is to ignore what is perhaps the bulkiest and most expensive problem facing the nuclear industry – what to do with old power reactors at the end of their lives. Although it was predicted that a significant proportion of the world's existing nuclear capacity would shut down by 2010 (Pollock, 1986), industry demands for the extension of operating licences, from typically 40 to as much as 60 years, will put off many such decisions until another day. Nevertheless, nearly 100 commercial reactors and many more small research reactors have already been shut down – although only 8 had been fully dismantled and 17 partially dismantled by 2006.

Decommissioning, rather than new plant construction, is likely to be the main growth area for the nuclear industry in many countries in the early 21st century. In its final stages, this involves the handling of very large pieces of radioactive material from the reactor core and associated cooling and heat transfer systems. Cooling water for concrete cutting saws, dust, cleaning solvents, etc. may all be potential agents for the spread of radioactivity and need to be carefully contained. The entire process was originally conceived as taking place in three stages:

1. Removal of fuel and 'mothballing' of the power plant, virtually intact but under 'care and maintenance';
2. dismantling and removal of all structures apart from primary containment around the reactor; and
3. dismantling and removal of remaining structures, with the eventual aim of returning the site to unrestricted use. This stage was assumed to be delayed in most cases by up to 100 years to allow radioactivity to decay, thereby reducing the cost and complexity of the final dismantling.

In practice, a different range of end-of-life options have been explored. For example, the International Atomic Energy Agency (IAEA) now recognises three alternative decommissioning strategies: De-con (immediate dismantling and removal); Safestore (placing the facility under long-term 'care and maintenance', with a view to later dismantling); and Entombment (encasing much of the radioactive material in concrete and/or under a landscaped mound, without specific plans for later removal).

In Britain, prior to the establishment of the Nuclear Decommissioning Authority (NDA) in 2005, the preferred strategy was Safestore, with the emphasis on extending the period of care and maintenance, following completion of the initial fuel-removal phase (usually within a few years of shutdown). The second phase of Safestore was planned to include the retrieval and packaging of radioactive wastes produced during operations, decontamination, dismantling and demolition of fuel-cooling ponds and removal of most non-radioactive plants, such as turbine halls. More recently, UK decommissioning policy under the NDA has swung back towards full decommissioning and clean up of sites, within as little as 25 years from shutdown, if possible.

Much of the attraction of the Safestore and Entombment options lies with deferring the problems and costs of large-scale waste disposal for future generations to handle, although the reduction in residual radioactivity (due to decay of short-lived isotopes created by irradiation of structural metals) will also reduce health and safety hazards during eventual dismantling. Some technological improvements may also be anticipated in the future. Set against this is the apparent irresponsibility of transferring liabilities to future generations, as well as the foregoing of the skills and experience of site operating staff.

Since few large reactors have yet been fully decommissioned following a typical commercial operating life (30 years or more), most experience to date has been obtained on early prototype reactors, where costs were expected to be exceptional. For example, the first British reactor to be fully decommissioned (the 33 MW Windscale AGR) is likely to cost at least £80 million ($130 million). In the United States, the costs of the demonstration tests for decommissioning the pioneering power plant at Shippingport, Pennsylvania, have never been fully released, and other American examples (e.g. the barely used plant at Shoreham, NY) are hardly representative. The full commercial costs of decommissioning therefore remain estimates at best. Previous assessments range from 10% to 100% of the original reactor cost (Pollock, 1986), i.e. at least US$100 million (£60 million) for a large commercial reactor. Over the lifetime of a power plant, decommissioning is expected to add about 5% to electricity costs, since many of these expenses may be discounted up to 100 years into the future.

However, the historic liabilities of early reactor types and their associated waste management facilities may be more substantial. When the NDA was set up in 2005, its liabilities were estimated at £56 billion, with much of this cost attributed to the problems of dealing with 'legacy' decommissioning and waste management. The NDA's 2006 published strategy revised these costs upwards to £72 billion. Worldwide, the nuclear industry argues that the relatively voluminous and difficult wastes generated where nuclear power programmes arose out of military programmes represent a special case, not entirely attributable to the cost of modern nuclear power generation.

In the United Kingdom, Berkeley and Hunterston 'A', two full-sized Magnox stations, had reached the care and maintenance phase of decommissioning by 2006, so decisions on future decommissioning policy are required imminently. Whatever the strategy chosen, the pressure will soon intensify to find storage space for spent fuel and repositories for separated waste.

Radioactive waste management

Compared with the problems of waste disposal and pollution from other parts of the energy sector (and industry in general), the relatively modest bulk of waste residues from nuclear power was originally considered to be a key virtue of nuclear power. From its early days until the 1960s, ocean dumping of low-level wastes (LLW) and the prospect of technological solutions to disposal of higher-level wastes was of minor concern. However, from the mid-1970s to the present day, mounting public concern and independent criticism of the nuclear industry has revealed waste management as its Achilles' heel. Much of the world's more dangerous radioactive waste (i.e. spent reactor fuel) is still in interim storage, mostly at the reactor sites themselves. While many countries have explored or supported deep geologic disposal as the best method for isolating highly radioactive, long-lived waste, no country has yet commissioned such a repository, although several have opened central interim stores for used reactor fuel and high-level wastes (HLW). Sweden and Finland are arguably the most advanced in their plans, promising operational repositories 'some time after 2010', but most other countries have vague timetables beyond 2020 or even 2030.

US government plans for a US$10 billion repository in a remote but geologically active site at Yucca Mountain, Nevada, originally planned for 2010, have been delayed until at least 2015 by legal challenges. In Britain, Nirex, the waste management company established in the wake of the Flowers Report (RCEP, 1976), initially planned to begin storing separated HLW in an underground repository at Sellafield by 2010, but this policy was later scaled back to consider only the much larger volumes of intermediate-level waste (ILW, e.g. reactor components, discarded fuel cans) and LLW (contaminated clothing, filters, etc.). In the mid-1990s, Nirex's plans to construct an underground rock laboratory at Sellafield for testing storage methods were held up by a public enquiry. After a string of government reports and consultations on radioactive waste management policy between 1999 and 2002, responsibility for the UK's nuclear legacy finally passed to the NDA, set up in 2005 as a non-departmental public body. As of 2006, there are no long-term arrangements for the management of HLW or ILW in the United Kingdom. Such wastes will be 'managed on an interim basis on sites managed by the NDA, possibly for several decades'. As Chapter 9 reports in detail, since 2003, the UK Committee on Radioactive Waste Management (CoRWM) has been reviewing the options for managing radioactive waste over the long term, and it will report later in 2006 on a recommended strategy.

Following the 1983 international moratorium on dumping LLW at sea, several countries have buried these wastes in shallow concrete-lined trenches. Britain has two LLW facilities, at Drigg near Sellafield, and a smaller one at Dounreay, but plans for further shallow LLW disposal sites were abandoned in the 1980s following vigorous local opposition. Increasingly strict regulation worldwide has produced a trend towards interim surface storage of nearly all radioactive wastes near their point of origin, with deep repositories held out as the long-term solution.

Plutonium and the fast reactor

The 'fast neutron', 'fast breeder' or simply 'fast' reactor is a type designed to produce more fuel while it is generating power. The reactor core requires the use of a more highly enriched uranium fuel, in order to sustain a chain reaction based on fast neutrons alone. This is surrounded by a 'blanket' layer of non-enriched uranium, which captures neutrons and is transformed into plutonium. The plutonium may then be extracted by reprocessing and used as fuel in another fast reactor, producing yet more plutonium – hence the term 'breeder'. This would obviate the need for expensive uranium enrichment and extend existing uranium reserves for many hundreds of years (compared with about 60–100 years, depending on the rate of use, for economically extractable uranium using conventional fission reactors).

However, the fast reactor philosophy was conceived at a time when the nuclear industry worldwide was expected to expand, thereby causing shortages of uranium ore. This was indeed the case in the mid-1970s, but the worldwide slowdown in reactor orders saw uranium prices drop well below $50/kg in the 1980s and 1990s. Despite a recent recovery, uranium prices are expected to remain around the range $50–100/kg until at least 2020, compared with the estimated $300/kg that would be required to justify the higher capital and fuel cycle costs of fast reactors.

Reprocessing has proved more difficult and expensive than uranium enrichment, and shortages of uranium are not forecast for decades to come. Furthermore, the fast reactor itself has proved technically complicated, with an intensely hot core subjecting the materials around it to high fluxes of both heat and neutrons. A molten sodium primary coolant is required to remove heat by conduction as well as by convection, but the presence of water as a secondary coolant means there must be strict safety measures, since sodium metal liberates hydrogen and explodes on contact with water. Although a number of large prototype fast reactors have been built worldwide, there is little prospect of commercial power production in the foreseeable future.

The UK's experimental 14 MW Dounreay Fast Reactor (started in 1955) and its successor, the 250 MW Prototype Fast Reactor (started in 1966), were deliberately sited on the remote north coast of Scotland due to uncertainties about the behaviour of such concentrated reactors. Despite much-delayed plans to construct a full-scale 1300 MW commercial fast reactor (for which Dounreay would have been a curious choice, being 70 miles from Britain's most northerly city, Inverness) the British fast reactor programme was scaled down from 1989 onwards. By this time, more than £3.5 billion ($5.6 billion) had been spent on research and development (Flood, 1988), but the Prototype Fast Reactor operated only intermittently from 1974 to 1994. Germany pulled out of its own $5 billion Kalkar fast reactor project in 1991, and the troubled and unreliable French 1240 MW Superphenix reactor was closed prematurely in 1998 after costs of around $10 billion. Both the joint European fast reactor research programme and similar Japanese plans to commercialise fast reactors have been repeatedly delayed.

Unfortunately for the nuclear power industry, there are unshakeable links between civilian and military plutonium stocks. For decades, governments and utilities alike maintained that the two could be kept functionally separate, but this distinction became very blurred in the post–Cold War 1990s. A fist-sized chunk of plutonium is enough for a terrorists' bomb – the bomb dropped on Nagasaki used only 10 kg – so the prospect of 50-tonne loads being moved around the world presents immense accounting and security problems, which may yet limit the prospects for future growth (Patterson, 1984; Beck, 1994). The ^{240}Pu and ^{241}Pu found in spent power reactor fuel (due to further neutron capture) may not be such good bomb material as 'weapons-grade' ^{239}Pu – it is hard to predict the size of the 'bang' – but this would not matter to terrorists or maverick governments who wished only to threaten others. It is not even necessary to obtain the critical mass of plutonium required for a bomb – atmospheric dispersal of just one kilogram of plutonium oxide using conventional explosives could contaminate the entire centre of a city with dangerous levels of extremely long-lived radioactivity (Miller, 1992). As the classic Flowers Report warned, 'the dangers of the creation of plutonium in large quantities in conditions of increasing world unrest are genuine and serious. We should not rely for energy supply on a process that produces such a hazardous substance as plutonium unless there is no reasonable alternative' (RCEP, 1976).

Nuclear fusion

Since it seems to be some way off providing anything approaching a commercial power reactor, fusion deserves only a brief mention here for the sake of completeness. Like fission, the thermonuclear fusion of small, light nuclei can release enormous amounts of energy – as utilised in the hydrogen bomb. However, the prospect of exploiting controlled nuclear fusion for electric power production remains a dream, forever 30 or 40 years distant. Despite the expenditure of at least £500 million ($800 million) on UK research since the early 1960s (Flood, 1988) and as much as US$20 billion worldwide (Beck, 1994), a commercial-scale power plant is unlikely to be constructed until at least 2040. Fusion has little chance of being a significant source of power before 2100.

Like the fast reactor, the engineering problems are substantial – in this case, the confinement of the heavy hydrogen isotopes deuterium and tritium within a toroidal (tyre-shaped) magnetic 'bottle' at a temperature of millions of degrees Celsius. Unlike the fast reactor, fusion offers the prospect of greatly reduced volume and intensity of radioactive waste, limited mainly to routine replacement of neutron-irradiated parts of the confinement vessel. The nuclear fuel is also (theoretically) cheap and abundant, with decreased environmental impact at all stages of the fuel cycle. An expensive research subject, fusion power nevertheless excites visionaries, and there has been enthusiasm for a range of novel approaches, including so-called cold fusion. It may be that eventually fusion will provide a useful energy source, for example, to power space-craft, but as far as terrestrial applications are concerned, for the foreseeable future, the sun seems to be a more realistic source of fusion energy.

References

Beck, P. (1994) *Prospects and Strategies for Nuclear Power: global boon or dangerous diversion?* Royal Institute for International Affairs, London/Earthscan Publ., London. 118 pp.

Flood, M. (1988) *The End of the Nuclear Dream: the UKAEA and its role in nuclear research and development.* Friends of the Earth Trust, London. 84 pp.

McKay, H.A.C. (1984) *The Making of the Atomic Age.* Oxford University Press/Oxford Paperbacks, Oxford. 153 pp.

Miller, G.T. (1992) *Living in the Environment.* 7th edition. Wadsworth, Belmont, California. 705 pp.

Patterson, W.C. (1984) *The Plutonium Business.* Paladin, London. 272 pp.

Pollock, C. (1986) *Decommissioning: nuclear power's missing link.* Worldwatch Paper 69, Worldwatch Institute, Washington, DC. 54 pp.

RCEP (1976) *Nuclear Power and the Environment.* Royal Commission on Environmental Pollution, 6th Report, Sept. 1976. HMSO, London. 237 pp.

Further reading

Berkhout, F. and H. Feiveson (1993) Securing nuclear materials in a changing world. *Ann. Rev. Energy Environ.* **18**, 631–665.

Bugl, J. (1992) Role of nuclear energy in reducing CO_2 emissions and requisite measures involved. In: *Technologies for a Greenhouse-Constrained Society* (M.A. Kuliasha *et al.*, eds.). Lewis Publ., Boca Raton, Florida. pp. 669–689.

Forsberg, C.W. and A.M. Weinberg (1990) Advanced reactors, passive safety and acceptance of nuclear energy. *Ann. Rev. Energy Environ.* **15**, 133–152.

Gowing, M. and L. Arnold (1974) *Independence and Deterrence: Britain and atomic energy 1945–1952.* Vol I – *Policy Making,* 464pp. Vol. II – *Policy Execution,* 559 pp. Macmillan, London.

Grimston, M.C. and P. Beck (2002) *Double or Quits: the future of civil nuclear energy.* Royal Institute for International Affairs, London/Earthscan Publ., London. 224 pp.

Kaijser, A. (1992) Redirecting power: Swedish nuclear power policies in historical perspective. *Ann. Rev. Energy Environ.* **17**, 437–462.

Keepin, B. and G. Kats (1988a) Global warming (Letter). *Science* **241**, 1027.

Keepin, B. and G. Kats (1988b) Greenhouse warming: comparative analysis of nuclear and efficiency abatement strategies. *Energy Policy* **16**, 538–561.

Nuttall, W.J. (2005) *Nuclear Renaissance: technologies and policies for the future of nuclear power.* Institute of Physics Publishing, Bristol. 322 pp.

Patterson, W.C. (1976) *Nuclear Power.* 2nd edition (1983). Penguin Books, London. 304 pp.

2
A Concise History of the Nuclear Industry Worldwide

Jonathan Scurlock

At the start of 2006, there were 441 nuclear power reactors in operation around the world, with a combined electrical capacity of about 370 gigawatts (GW), producing around 2600 terawatt-hours (TWh) per year or roughly 16% of world electricity demand. About one-third of this capacity is found in North America, and one-third in the European Union. Having outlined the basic science of nuclear energy in the previous chapter, the following sets the scene for the subsequent discussion of whether or not the nuclear contribution to electricity supply should be expanded, by describing briefly how the existing civil nuclear programme unfolded around the world.

Early years – war and peace

Launched in the middle of the Second World War, the Manhattan Project was one of the largest scientific undertakings of the 20th century, as an international team of experts collaborated with military and industrial engineers in the United States in a race to develop an atomic bomb before Nazi Germany. The first experimental nuclear reactor to 'go critical' without an external source of neutrons was constructed in Chicago in the late 1942 – but the Americans withdrew behind a veil of military secrecy in 1946, and the international effort was succeeded by national programmes.

The earliest large nuclear reactors built in the USA, Britain, USSR and China were all designed to make weapons-grade plutonium for atomic bombs. These *atomic piles* comprised simply of piles of graphite blocks into which uranium reactor fuel was loaded for transformation into plutonium. They were water-cooled (as at Hanford, Washington state) or air-cooled (as at Oak Ridge, Tennessee, and later at Windscale, UK).

However, the rather injudicious use of air for cooling the hot graphite led to the world's first major reactor accident – the Windscale fire of 1957.

The first electricity to be generated by nuclear power, in 1951, actually came from a small breeder reactor in Idaho, USA, named EBR-1. But the US nuclear research programme was directed less at electricity gener- ation and more towards propulsion for submarines, then seen as a stra- tegically important application of nuclear power. A nuclear submarine could remain underwater for months at a time without requiring air for its engines. Powered by a new compact reactor design, consisting of a small core of enriched uranium fuel with pressurised water serving a dual role as a moderator and coolant, the *USS Nautilus* entered service in 1954.

Meanwhile, the United Kingdom and France based their civilian power reactor designs on the earlier plutonium-producing reactors, and in 1956, Britain put into operation the first prototype nuclear power station, even- tually comprising four 50 MW reactors at Calder Hall. Like the nearby Windscale plutonium production piles, Calder Hall used graphite blocks as a moderator, but was cooled by carbon dioxide instead of air. It is no secret that it was intended for both electricity and plutonium produc- tion, and estimates of the cost of nuclear electricity from the derivative Magnox reactors were later revised by the UK Atomic Energy Authority to include a credit for creation of plutonium (Flood, 1988). In the same year, the French also started up an air-cooled, graphite-moderated 40 MW reactor for plutonium and power production at Marcoule.

The first large American power reactor, which began operating in 1957 at Shippingport, Pennsylvania, was a 60 MW unit hastily modified from the US military submarine design, and originally destined for a nuclear aircraft carrier. Thus a compact American configuration, intended for the restricted space onboard a submarine, evolved into the most commonly used reactor types today, the pressurised water reactor (PWR) and the related boiling water reactor (BWR). Water is held under pressure in order to keep it below or close to boiling point at the reactor operating tem- perature of 300°C, so the reactor core is surrounded by an immensely strong pressure vessel (Figure 2.1a). This must be capable of withstanding 150 times atmospheric pressure in the case of the PWR, and somewhat less in the case of the BWR, where the water is allowed to boil. PWRs and BWRs are often referred to jointly as *light water reactors* (LWR), distinguishing them from reactors which use heavy water as the moderator. A significant weakness of this design, posing a major environmental risk, stems from the containment of the *primary* water circuit under high pressure. Any failure leading to a release of steam may spread radioactive contamin- ation over a large area. LWRs are also subject to engineering problems in the heat exchangers between the primary and the *secondary* cooling water. A substantial literature has grown up around the safety of LWRs, their performance under both real and hypothetical 'loss-of-coolant' incidents, and the need for emergency core cooling systems (Patterson, 1976).

Scaling up – the 1960s

As governments sought to support an electricity generation industry based around nuclear power, the size of power reactors grew. But energy was cheap, and there was little incentive for industry to invest in nuclear technology without heavy government subsidies. American electricity utilities refused to participate at first, on the grounds of cost, risk and the availability of cheap oil and abundant coal. The US government responded by building a series of demonstration reactors using different technologies. Most of these performed poorly, and only the PWR and BWR emerged as leading contenders.

Beginning with a 500 MW BWR at Oyster Creek, New Jersey, in 1963, a string of fixed-price (and often loss-making) commercial contracts for nuclear power stations were let by competitors General Electric and Westinghouse. Losses of up to $1000 million per plant are thought to have been sustained by the manufacturers in their determination to build up the market (Bupp and Derian, 1978). But in the 'Great Bandwagon' of orders from American utility companies, some 40 GW (44 plants) were ordered in 1966–67 alone.

Meanwhile the first British commercial reactors were based on the early Calder Hall and Chapelcross dual-purpose electricity/plutonium reactors. The *Magnox* series was named after the alloy cladding around the fuel rods, and the reactor building programme was expanded in response to strategic energy concerns following the 1957 Suez Crisis. The first Magnox power station entered service in 1962 at Berkeley, Gloucestershire, and, barely pausing to consult the Central Electricity Authority (later the Central Electricity Generating Board – CEGB), the government and UK Atomic Energy Authority pressed ahead with a programme of eventually nine such stations – unfortunately, with an abundance of variants as successive orders went to an assortment of different engineering consortia (Patterson, 1985).

The second wave of the British nuclear power programme was subject to a greater degree of debate about reactor types. The CEGB favoured North American designs over those of the UK Atomic Energy Authority, but the government opted for a refinement of the Magnox design – more compact, with a higher operating temperature, running on enriched

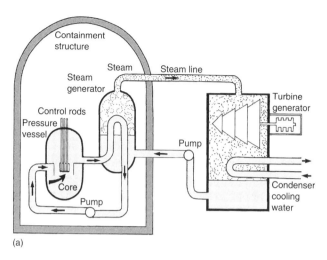

(a)

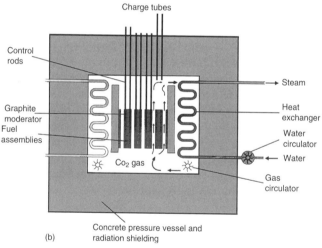

(b)

Figure 2.1 Principal features of two significant types of nuclear power reactor: (a) generalised PWR, e.g. US Westinghouse design, British modified Westinghouse reactor at Sizewell 'B', French Framatome variant, and others; (b) AGR, developed and used exclusively in the United Kingdom. The Magnox reactor is similar, but with a more massive core. Note the highly compact core of the PWR, originally intended for marine applications, and the distinction between *primary* containment or shielding immediately around the core, and *secondary* containment around the entire reactor system. By comparison, the core of the AGR is much larger, but the pressure vessel and containment are integral to the structure of the whole system. In both cases, note the essential role of the heat exchangers between primary and secondary cooling circuits and the need for active circulation of the primary coolant.

uranium oxide fuel – the advanced gas-cooled reactor (AGR). Following an appraisal of projected costs for the Dungeness 'B' power station, the British Minister of Power announced a programme of AGRs in 1965, but Dungeness 'B' was an inauspicious start: it took 17 years to complete and set records for poor output thereafter (Patterson, 1985) (Figure 2.1b).

Strategic growth – the early 1970s

With the first OPEC oil price shock of 1973, there came a further rush of orders for the nuclear industry throughout the industrialised world, based on the notion of comparative independence of energy supply. The world's uranium resources were seen to be more widely distributed that oil (especially within industrialised nations such as Canada and Australia), and the quantities of uranium required were small compared with those for coal, more easily traded and therefore 'strike-proof' (note that the international trade in coal in the 1970s was a fraction of what it is today).

In France, the nuclear industry switched from British-style gas-cooled graphite-moderated reactors to US-style PWRs at the end of the 1960s, at which time France was highly dependent on imported oil and gas for power generation. In response to the oil crisis, state-controlled Electricité de France (EdF) enacted a massive expansion of nuclear power throughout the 1970s, based on series production of Westinghouse-type PWRs. Construction of 34×900 MW units, then 22×1300 MW units, took nuclear power generation from 7% of total electricity in 1973 to 20% in 1980 and 78% by 1994. Construction times for the standardised power stations were as short as 5 years, and the French nuclear programme as a whole was widely admired by nuclear advocates in other countries – although, like elsewhere, it has remained firmly under state control with very limited public participation in decision-making.

At the same time, Belgium, Germany and Sweden were all developing significant nuclear programmes, based mainly on LWRs. Meanwhile, Canada was going it alone with the novel CANDU reactor, a design using non-enriched uranium (of which the Canadians had plenty) and heavy water as the moderator. The first commercial CANDUs came on stream in 1971 at Pickering, near Toronto (Patterson, 1976). These were significant for being the first reactors designed specifically for power production (with no military or strategic overtones) and for incorporating a number of passive safety features. Unfortunately, cost over-runs on heavy water production seriously tarnished the profitability and image of the CANDU reactors, despite their reliability and high capacity factors.

Ironically, the CANDU's ability to be refuelled continuously under load means that it can also be used for production of weapons-grade plutonium (^{239}Pu) – a fact that caused great embarrassment to the Canadians in 1974 when India exploded a device containing plutonium from a CANDU-type research reactor (Patterson, 1984).

Overall, at least 20 different nuclear reactor concepts had received serious attention by the mid-1970s, but only four types have been used for commercial electricity production (five if sodium-cooled fast breeders are included): (1) light water (PWRs, BWRs); (2) heavy water (e.g. CANDU); (3) steam-cooled graphite reactors (e.g. Russian RBMK) and (4) gas-cooled graphite (e.g. Magnox, AGRs). However, LWRs predominate, making up more than 80% of reactors worldwide today (Table 2.1).

Reaching a turning point

The nuclear fervour of the early 1970s was short-lived. The worldwide economic recession which followed the 'oil crisis' inflated the cost of capital-intensive nuclear construction projects, while reducing present and future projections of energy demand. Although about 40 reactors were ordered in the United States in each of 1973 and 1974, barely one single power station has been ordered and completed since, and no new

Table 2.1 Major nuclear reactor types and their approximate installed capacity worldwide (various sources, 2005/2006)

	Total GW	LWR	CANDU	Steam-graphite	Gas-graphite
North America	110	98	12	—	—
Latin America (Brazil/Argentina)	3	2	1	—	—
Europe	139	128	—	—	11
Former Soviet Union	44	33	—	11	—
Africa (South Africa)	2	2	—	—	—
Middle East/S. Asia (India/Pakistan)	3	—	3	—	—
Far East, S.E. Asia and Pacific	67	64	3	—	—
	368	327	19	11	11

Notes: Figures are rounded in gigawatts (GW) to denote the approximate capacity of working reactors (note that many sources differ between 'operable' reactors, reactors undergoing commissioning and reactors under construction, as well as 'design' and 'effective' installed capacity). LWR = light water reactors (PWRs, BWRs); CANDU = Canadian-type heavy-water moderated reactors; Steam-graphite = steam-cooled, graphite-moderated reactors (e.g. Russian RBMK type); Gas-graphite = gas-cooled, graphite-moderated reactors (e.g. British Magnox and AGRs) orders were placed after 1978. Some US reactors were abandoned when more than 90% complete.

The projections previously made by the nuclear industry about its own prospects were ambitious, perhaps even arrogant (Patterson, 1985), and they had to be scaled back. In the early 1970s, the United States was predicted to have around 1000 nuclear power stations in operation by the year 2000 – instead it managed barely more than 100. Worldwide, it was estimated that there would be 4000 GW of capacity worldwide by the millennium (i.e. about 4000 large power plants) – the reality fell short of 400 GW, less than 10% of predictions.

In Britain, a suddenly enthusiastic CEGB told a Parliamentary committee on science and technology in 1973 of their plans to order 32 large (1200–1300 MW) PWRs within 10 years (the oil crisis and a work-to-rule by British coal miners probably had something to do with this). But the CEGB's preference for LWRs did not prevent the fiasco of the Steam Generating Heavy Water Reactor, six of which were ordered by the UK Department of Energy in 1974, only to be abandoned in 1978 following further economic recession – by which time £145 million had already been spent on development (Patterson, 1985). Meanwhile, in 1975, in its evidence to the Royal Commission on Environmental Pollution, the UK Atomic Energy Authority announced a 'reference programme' in which British nuclear capacity would grow from 5 GW in 1975 to 104 GW (more than 100 large power stations) by the turn of the century. Of this future capacity, 33 GW was to consist of plutonium-fuelled fast breeder reactors using liquid sodium coolant (Patterson, 1985). The reality for the British nuclear industry has been exceptionally modest by comparison – about 13 GW installed capacity in 2000, with just two AGRs completed (at Torness and Heysham) and one PWR (Sizewell 'B') since the mid-1970s.

Economics and safety matters

In the late 1970s, some of the real financial costs of nuclear power came to light in a study by two American analysts from Harvard Business School (Bupp and Derian, 1978). Prior to that, nuclear economics had been a kind of inspired guesswork based on anticipated high capital costs and low operating and maintenance costs, leading to the fixed-price (and loss-making) contracts drawn up by US power station constructors. Later nuclear power station orders proved to be expensive mistakes for their customers, the US utility companies. For example, the Washington Public Power Supply System defaulted on over US$2 billion (£1.3 billion) of bonds after cancelling four power plants. Other plants such as Shoreham, New

York, and Seabrook, New Hampshire, cost around $5 billion each (£3 billion) due to repeated delays in construction and licensing, and the former never actually produced any electricity (Morone and Woodhouse, 1989). Meanwhile, in France, EdF's nuclear construction programme and low electricity tariffs had resulted in accumulated debts of £29 billion by the end of the 1980s (Miller, 1992) – it was clear that the French enthusiasm for nuclear was principally strategic, not economic.

In the late 1980s, British government plans for privatisation of the electricity supply industry included the CEGB's stock of Magnox and AGR plants, as well as the single PWR. But the exposure of the economics of nuclear power in the UK showed that neither the ageing Magnox stations nor the AGRs were financially viable, based on poor collective performance and a high degree of uncertainty over the cost of future decommissioning (Parker and Surrey, 1995). Nuclear power was abruptly withdrawn from the 1989 privatisation, with the government-owned companies Nuclear Electric (in England and Wales) and Scottish Nuclear (in Scotland) retaining the stock of commercial nuclear power stations. Electricity from nuclear generation was subsidised from 1989 onwards by payment of a premium price through the Non-Fossil Fuel Obligation (NFFO), financed by a levy on electricity tariffs, initially set at 11%. The NFFO, ostensibly a scheme to promote diversity of supply and control CO_2 emissions, raised about £1.2 billion a year ($1.9 billion) during the early 1990s, of which 92–98% was paid out to Nuclear Electric (the remainder going to a variety of commercial renewable energy projects). An official review of nuclear power policy carried out in 1994–95 concluded that, under improved management, the more modern stations were suitable candidates for privatisation, after all (DTI, 1995). However, it also stated that the provision of public funds for new power plant construction was not warranted on the grounds of economics, emissions abatement, diversity or wider economic benefits. This decision finally brought to an end a long era of continuous forward planning of UK nuclear reactor construction.

Despite the concerns raised by the British Royal Commission report (RCEP, 1976) and others, the safety of nuclear power did not really become an issue until 1979 – it was previously dismissed on the grounds of statistical analysis of component failures, each with a low probability. Although there was no breach of the pressure vessel or direct loss of life, the March 1979 loss of coolant and partial core meltdown in one of two PWRs at Three Mile Island near Harrisburg, Pennsylvania, was a severe setback for nuclear power worldwide, leading to moratoria on future construction in Italy, Belgium, Sweden and elsewhere. Costing up to

US$4 billion (£2.5 billion) and involving the voluntary evacuation of over 100,000 people in the resulting media confusion (a mandatory evacuation was never called), the Three Mile Island accident brought the US nuclear power programme to a standstill – but the slowdown in power station orders had begun several years earlier, for economic reasons.

By 1986, oil prices had fallen, coal was becoming available as an internationally traded commodity and estimates of natural gas reserves were climbing. The economic prospects for nuclear power were therefore looking worse still when, in April, a large leak of radioactivity was detected in the atmosphere over Sweden. One of four 1000 MW steam-cooled graphite-moderated RBMK reactors at the Chernobyl complex near Kiev, Ukraine, had suffered a steam explosion, graphite fire and core meltdown two days earlier, but the then-secretive USSR government had initially tried to cover up the accident. The Chernobyl disaster resulted in at least 50 immediate fatalities and several thousand consequential deaths, as well as the evacuation of 135,000 people, resettlement of more than twice this number, and direct financial losses of at least $3 billion (£2 billion). The eventual cost to the national economies of the area (including loss of agricultural land and construction of permanent containment around the site) has been estimated at several hundred billion dollars (Miller, 1992).

There is little doubt that Chernobyl contributed to the 1989 fall of the Berlin Wall and the 1991 demise of the former Soviet Union. It also marked a symbolic opening of the world nuclear industry to careful scrutiny by economists, investors and the public. Nuclear technology experts from the United States, Japan and Western Europe have been visiting and advising their Eastern European and Asian counterparts ever since, in the knowledge that if another reactor melts down or leaks badly anywhere in the world, their industry faces almost certain shutdown.

Growth in world nuclear capacity has slowed significantly in recent years. Between 1999 and 2004 there was a net increase in generating capacity of only 14 GW, and the number of operating reactors has barely increased, as larger units replace smaller ones (WNA, 2006). In the absence of government support, liberalised energy markets in OECD countries are presently choosing not to build new nuclear plants. Only about 1–2% of the world's new power plants are nuclear, and today most of the reactors under construction, on order or planned are in Asian countries such as China and India (and to a lesser extent, Japan). Nevertheless, pressure has been mounting for a reassessment of the nuclear option in the United States and Europe, in part because of the potential role that nuclear power can play in response to climate change. The prospects for expansion in Asia are discussed in a later chapter. But wherever expansion is considered, as this brief history has illustrated, there will be many technical and financial obstacles that the industry has to overcome, as well as a range of strategic security issues, some of which are discussed later in this book. However, probably the most obvious problem that any nuclear renaissance will have to face is the widespread public opposition that has emerged in many countries around the world, an issue taken up in the next chapter.

References

Bupp, I.C. and J.C. Derian (1978) *Light Water: how the nuclear dream dissolved.* Basic Books, New York. 241 pp.

DTI (1995) *The Prospects for Nuclear Power: conclusions of the government's nuclear review*, Cm 2860. Department of Trade and Industry, London/HMSO, Norwich.

Flood, M. (1988) *The End of the Nuclear Dream: the UKAEA and its role in nuclear research and development.* Friends of the Earth Trust, London. 84 pp.

Miller, G.T. (1992) *Living in the Environment*, 7th edition. Wadsworth, Belmont, California. 705 pp.

Morone, J.G. and E.J. Woodhouse (1989) *The Demise of Nuclear Energy?: lessons for democratic control of technology.* Yale University Press, New Haven. 172 pp.

Parker, M. and J. Surrey (1995) Contrasting British policies for coal and nuclear power, 1979–92. *Energy Policy* **23**, 821–850.

Patterson, W.C. (1976) *Nuclear Power*. 2nd edition (1983). Penguin Books, London. 304 pp.

Patterson, W.C. (1984) *The Plutonium Business*. Paladin, London. 272 pp.

Patterson, W.C. (1985) *Going Critical: an unofficial history of British nuclear power.* Paladin, London. 184 pp. also available at: www.waltpatterson.org (accessed 5th April 2006).

RCEP (1976) *Nuclear Power and the Environment.* Royal Commission on Environmental Pollution, 6th Report, Sept. 1976. HMSO, London. 237 pp.

WNA (2006) World Nuclear Association web site http://www.world-nuclear.org, accessed 29th March 2006.

3
Opposition to Nuclear Power: A Brief History

Horace Herring

Twentieth-century nuclear visions

With the publication of *The Interpretation of Radium* (1909), Frederick Soddy, who was a leading nuclear physicist as well as a popular author, began the process of mythicizing atomic energy. It was presented as an inexhaustible supply of power, which could be used to transform society, heralding the possibility of an atomic utopia. His views on atomic energy had a great impact on the public, and inspired H. G. Wells to write his famous science fiction novel *The World Set Free* (subtitled *A Story of Mankind*) published in 1914. Wells gave the world the first vision of what he called the 'atomic bomb'. He presents a very bleak picture of the horrors of atomic warfare with hundreds of cities destroyed but later the atom is used for peaceful purposes and it concludes with a cultural renaissance spawned by atomic power. This novel was highly influential in giving us conflicting images of atomic power, which were utilized by later science fiction writers (Herring, 2005).

The roots of nuclear euphoria (and ambiguity) actually date back further, to the discovery of the phenomenon of radioactivity in 1896, due to X-ray emissions from radium. This led to a 'radium craze' among the public, with radioactivity being seen as a scientific miracle with a wide range of positive health effects. Newspaper coverage in the early twentieth century was strongly positive, even though it was acknowledged that exposure to radium causes burns and eventual death. Scientists who died were considered 'martyrs to science', and the press, scientists and industry all promoted the view that the benefits of radium, particularly for 'curing' cancer, strongly outweighed any hazards.

The discovery of nuclear fission produced by U-235 in 1938 unleashed a torrent of similar imagery: nuclear-powered planes and automobiles would whisk us effortlessly around the globe, while unlimited nuclear electricity powered underground cities, farms and factories. However the dropping of the two atomic bombs on Japan in 1945 caused intense public debate in the United States and elsewhere on the morality of using atomic weapons, and fear and anxiety about the consequences of atomic warfare. This public anxiety faded in a year or two to be largely replaced with a post-war nuclear euphoria and government optimism of 'the almost limitless beneficial applications of atomic energy'. But by the early 1950s the dream of atomic energy had stalled, despite popular enthusiasm. Some economists openly expressed doubts as to its economic feasibility and practicality, and there was little immediate prospect and much disagreement on any likely timetable for the construction of a commercial nuclear power station.

Atoms for peace

What revitalized the atomic dream and launched a commercial programme was the 'Atoms for Peace' speech by President Eisenhower on 8 December 1953 at the UN. It marked a major shift in US government atomic policy, ending the government monopoly on nuclear power, and rekindling the idea of a nuclear utopia. He urged that nuclear materials should be used for peaceful purposes and to 'provide abundant electrical energy in the power-starved areas of the world'. Once again the atomic visionaries rushed into print repeating the old 1940 predictions with a few new ones. There would be nuclear-powered planes, trains, ships and rockets; nuclear energy would genetically alter crops and preserve grains and fish; and nuclear reactors would generate very cheap electricity.

The most famous phrase of this era, and one that was to haunt the nuclear industry for evermore, was uttered by Lewis Strauss, the new chairman of the US Atomic Energy Commission (AEC), in his speech on 16 September 1954. Electric power from the atom, he said, could be available, in 'from five to fifteen years'. He then went on to give his vision of the atomic utopia, saying: 'It is not too much to expect that our children will enjoy in their homes electrical energy too cheap to meter, will know of great periodic regional famines in the world only as matters of history, will travel effortlessly over the seas and under them and through the air with a minimum of danger and at great speeds, and will experience a life span far longer than ours ... This is the forecast for an age of peace'. This nuclear utopianism went virtually unchallenged. For there was unquestioning support from the media, and unqualified endorsement by Congress and the administration.

After Eisenhower's Atoms for Peace speech there was rapid commercial development of nuclear power. Construction started on the first reactor in September 1954 at Shippingport, but it did not provide power until December 1957. The first commercial electric power from atomic energy in the United States came from a prototype reactor at West Milton, NY on 18 July 1955, which was greeted with much acclaim and an editorial in *Science* magazine.

The pace of commercial nuclear power development quickened from December 1963 when General Electric offered a fixed price (turnkey) contract to build the Oyster Creek power station. By the end of 1966 another 11 turnkey contracts were signed together with 16 other contracts. The rapid expansion of nuclear power in the United States in the 1960s was, according to nuclear historian Steven Cohn in his book *Too Cheap to Meter*, partly due to the public's buoyant faith in business and public leaders and their faith in the ability of science and technology to unambiguously solve social and economic problems. By the mid-1960s with a large nuclear programme underway the dream of atomic energy and an all-electric future seemed imminent.

Although an anti-nuclear movement gathered pace in the United States, with reactor safety being a key issue, in the late 1960s and early 1970s, the nuclear vision remained undimmed in the eyes of its advocates. In 1971 Glenn Seaborg, the head of the AEC from 1961–71, and William Corliss published their book *Man and the Atom*, subtitled *Building a New World through Nuclear Technology*. In predictions for 2000 they forecast that nuclear power would be a phenomenal success, bringing unimagined benefits for the greater part of humanity. Seaborg believed that the future of civilization was in the hands of the nuclear scientists who formed the elite team that would build a new world through nuclear technology.

Atomic energy in Britain

In Britain, the 1955 White Paper *A Programme of Nuclear Energy*, on the first nuclear power programme, was greeted with acclaim by the press, which followed it up with glowing reports on the triumphal opening of Calder Hall by the Queen in October 1956 – over a year before the Shippingport plant of the United States was started up. After the Suez Crisis in 1956, the government took the opportunity to triple the expansion of the nuclear industry. Until the early 1970s there was little criticism of nuclear power, and it was restricted to specialist journals, elite newspapers (generally *The Times*), parliament and industry forums. A small group of parliamentary backbenchers, however, raised a series of issues, including reactor safety, the adequacy of plans for nuclear waste disposal, the dangers of low-level radiation and the absence of any integrated energy policy.

The public inquiries

Between 1956 and 1961 there were seven public inquiries into proposals to build Magnox stations. The first (commercial) Magnox, at Berkeley in 1955, was approved without an inquiry but from then on every Magnox, except Sizewell A in 1960, was subject to an inquiry lasting from two to five days. This compares to all five AGRs built without an inquiry in the 1960s. In fact there was a ten-year interval (1961–71) without an inquiry. The 1970s saw a return to nuclear inquires. In 1971 there were two: the first over Connah's Quay lasted 28 days, and the second over Portskewett for three days. Then in 1974 there was an eight-day inquiry over Torness followed by the marathon hundred-day inquiry over the Windscale THORP plan in 1977. Subsequently there were major inquiries into the proposed Sizewell B PWR (which was passed) and Hinkley Point C PWR (which was also passed, but not in the event built).

National press coverage of the earlier inquiries was mostly confined to short reports or letters, mostly on wildlife or amenity issues, with local press giving more space to opposition groups. Amenity issues were central to objectors' arguments and press coverage, and reflected the fact that amenity was a familiar reason for dissent. In the prevailing language of the 1950s amenity had a similar connotation to NIMBY ('Not in my back yard') reactions in more recent times. The building of large industrial plants in rural areas was considered a threat to England's 'green and pleasant land' and found expression in the House of Lords, the press and at the inquiry. The amenity issue was especially important in two location decisions, at Trawsfynydd within the Snowdonia National Park and at Dungeness due to the outstanding beauty of these areas and the efforts of naturalists in opposing the developments.

National Coal Board campaign

Most criticism in the 1960s was of the decisions made on the speed, scale and reactor choice of the nuclear programme, not of the nuclear project itself. In the early 1960s the UK Atomic Energy Authority (AEA) forecasts were quickly proved in several respects to be over-optimistic: the Magnox reactors took longer and cost more to build than expected, while the cost of power from coal fell. The appointment of Alf Robens as chairman of the National Coal Board (NCB) in 1960 heralded the start of the first national campaign against nuclear power in Britain. His ten-year campaign was ably assisted by Fritz Schumacher, the chief economist of the NCB. Robens and the Coal Board as a whole were not against the building of nuclear power stations, but wished to protect the coal industry from a too rapid a shut down. Schumacher, in contrast, was against any nuclear power and continually pointed out the finite nature of the non-renewable energy resources and the foolishness of abandoning one major source just because the other was claimed to be cheaper in the short term.

The years 1965–68 had two White Papers on *Fuel Policy*, in 1965 and 1967, accompanied by much internal fighting between the coal and nuclear industries, and extensive newspaper coverage on the (bleak) future for coal and (rosy) prospects for nuclear. Robens's campaign to protect the coal industry and to expose the subsidies given to nuclear power had the backing of many Labour MPs, especially those from mining constituencies. However the Labour government, under Harold Wilson, was very pro-nuclear and the established view was that the coal industry had no future and should be run down. The campaign reached a climax in 1967–68 over the proposed AGR nuclear power station near Hartlepool, south of Newcastle. Robens realized that if this station was built, on the very edge of the Durham coalfield, then the coal industry had a bleak future as a supplier of coal to the electricity industry.

Robens, along with the Select Committee on Science and Technology, repeatedly called for an independent financial investigation of the economics of nuclear power as he did not believe CEGB's claims that nuclear was cheaper than coal and complained bitterly about nuclear secrecy on costs. However the government had been persuaded, by a 1964 review, that nuclear power in the form of the AGR was the power station of the future. The result, however, was a foregone conclusion: the media and many MPs, such as Tony Benn, were solidly pro-nuclear and there were then few anti-nuclear critics in Britain. With the publication of the 1967 White Paper (which gave an enhanced role to nuclear and a reduced one to coal) together with the approval given to the Hartlepool nuclear station in 1968, Robens realized that the coal industry had suffered a serious defeat. His anti-nuclear campaign had been in vain. Both he and Schumacher left the NCB soon afterwards. In the 1970s Schumacher became a vocal spokesman for the anti-nuclear campaign, particularly after winning worldwide fame with the publication of his book *Small is Beautiful* (1973).

The anti-nuclear campaign in the 1970s

The year 1970 can be seen as a divide between the old conservation groups and the new environmental ones. It was the year of the founding of the UK branch of Friends of the Earth (FoE). It was also the year of the first local-level protest for a decade in Britain against the construction of a nuclear power station, and also of increasing media (and environmental press) attention to the nuclear opposition in the United States. By 1970 two influential US anti-nuclear books had reached Britain – Curtis and Hogan's *The Perils of the Peaceful Atom* (1969), and Gofman and Tamplin's *'Population Control' through Nuclear Pollution* (1970). The continual drip of articles in the national press on the dangers of low-level radiation may have inspired environmental writers. In the summer of 1970, Walt Patterson and Peter Bunyard, who were both to become prolific and well-known writers on nuclear issues, published their first critical articles on nuclear power in the newly emerging environmental press.

Initially this eco-press concentrated their criticism on the American LWRs – the British AGR reactors overall received a relatively favourable press. However wider anti-nuclear arguments were steadily introduced (from the USA), including the risks of low-level radiation, the link between civil and military uses, and the dangers of nuclear terrorism. However, unlike the USA, France or Germany, there was no mass campaign. Partly this was because no new stations were being proposed, and partly because there was little public discussion of nuclear affairs. The old style conservationists and new left were generally in favour of nuclear power, while most environmentalists were concerned with issues of global doom – the *Limits to Growth* and *Blueprint for Survival* debate. Few had thought about nuclear issues.

Early anti-nuclear activism reflected the style and tactics of existing conservation groups, mainly middle-class and middle-aged. The earliest protests, such as at Bradwell, involved letter writing and conventional political lobbying. The only demonstration was at Trawsfynydd, and that was by local people in favour of the nuclear reactor! Protest was initially small and scattered, but under the leadership first of the Conservation Society (ConSoc) and then of FoE, it attracted increasing support from a wide range of groups.

The first national campaign was started by ConSoc in 1974, but its conventional approach attracted little support from young environmentalists. By late 1975 Beryl Kemp, its organizer, reported that she was becoming disillusioned with the lack of progress through the 'constitutional' approach and wanted to use a new form of action, 'non-violent resistance action', which seemed so successful overseas. However it was not the tactics that were wrong, but the dullness of the anti-nuclear power campaign run by ConSoc. The anti-nuclear campaign did not attract many followers until *Undercurrents*, the leading environmental magazine, had its first issue on nuclear power in early 1975. This publicity for an anti-nuclear campaign was then harnessed by FoE, who were able to launch their own campaign in May 1975 five months before the *Daily Mirror* made nuclear power, and THORP in particular, the centre of national attention.

THORP and the saga of reprocessing

As noted in Chapter 1, the initial primary reason for reprocessing spent reactor fuel at Windscale (on the site that later became know as Sellafield) was to extract the plutonium for use in nuclear weapons, although it was also expected that it would be used in fast breeder reactors (FBRs). The commitment to reprocessing was further strengthened in the late 1960s by the failure of the British Magnox and AGR reactor designs to win export orders. The only commercial opportunity left was in the reprocessing market, and it was realized that a dedicated plant would have to be constructed if Britain were to become a major player in the expanding market for reprocessing services. Hence the decision in the early 1970s by British Nuclear Fuels (BNFL) to construct the Thermal Oxide Reprocessing Plant (THORP) at Windscale. Thus the support of the British government and media's early enthusiasm for THORP rested on the perception that reprocessing gave easy profits and large foreign earnings at essentially no risk. It was taken for granted that THORP would be approved through the local planning processes without recourse to public inquiry and with little public comment.

However from 1973 a series of accidents resulted in large discharges of radioactivity into the Irish Sea, and provided unwelcome publicity for Windscale's activities. Then in October 1975 a front page story on Britain's 'Nuclear Dustbin' in the *Daily Mirror* made THORP a household name. Despite rising public concern the government was keen to push THORP forward and tried to play down calls for an inquiry. However a U-turn came in late December 1976, after it emerged that the government had not been informed by BNFL of a further leak of radioactivity. After that it could no longer resist the clamour for a public inquiry, and this was held from June to October 1977 under Mr Justice Parker. This inquiry was not only the battleground but also the recruiting ground for the fledgling British anti-nuclear power movement. It also marked the turning point in the environment movement from the generally pro-nuclear stance of the old conservationists of the 1960s to the now preva-lent anti-nuclear views of modern environmental activism as typified by FoE and Greenpeace.

THORP constructed

Justice Parker's *Windscale Report* in early 1978 was in favour of THORP and totally rejected the arguments of environmentalists against the plant. It was overwhelmingly approved by parliament. It was a major project, and met with some delays. Construction was started in 1985 and it was not completed until 1992. However the energy world had changed greatly since THORP was approved. Its main strategic rationale in the 1970s – the provision of plutonium to fuel the FBR – had col-lapsed, as the FBR was no longer considered a commercially viable reac-tor. Since there was no demand for plutonium, reprocessing was now unnecessary, and its operation only aggravated the problems of nuclear waste disposal and nuclear proliferation. Instead of THORP being risk-less 'easy money' for Britain, it now appeared to many as a huge liability and this concerned the Treasury who became increasingly worried in the early 1990s about THORP's financial risks.

Before it could start operating, BNFL needed various regulatory authorizations for its discharges of radioactivity. The government car-ried out a public consultation exercise, prior to a parliamentary debate on the issue in June 1993 which ended with vote in favour. Following official approval for BNFL's application in December, Greenpeace imme-diately applied for a judicial review, but this found in favour of the min-isters on all relevant points. So in late March 1994, BNFL received the authorizations it had applied for two years previously, and the radio-active commissioning of THORP began.

The opposition in the 1990s to THORP was similar to that in the mid-1970s – loud but ineffectual. This time it was led by Greenpeace, rather than FoE, again supported by local groups such as CORE (Cumbrians Opposed to a Radioactive Environment). Greenpeace's efforts had little effect, as it could not use one of its most effective tactics, namely con-sumer boycott (which it was to do with great effect with Shell over Brent Spa in 1995). Also THORP never became a party political issue, with par-liament showing little interest during this period. Indeed the Conservative government was always confident of the support of the Labour party – THORP had been approved by the last Labour government. Then there was strong trade union support for THORP, as the Sellafield site was heavily unionized and BNFL had long been skilled at winning influence in the Labour party through the trade unions. Finally THORP was too local, arcane and complex an issue to sway whole electorates. The Conservative and Labour parties sensed that THORP carried few electoral risks, despite huge media coverage which played up fears of cancer and childhood leukaemia.

THORP update

Since its opening THORP has performed fitfully and has often shut down. Since April 2005 it has been closed following a leak of nitric acid containing 22 tonnes of dissolved uranium and plutonium which begun in July 2004 but was not noticed until the following April. Reprocessing has therefore had a slow and erratic beginning. The future looks no better, for BNFL plans to close THORP around 2010 – when it has fulfilled its current contracts. The end result of reprocessing has not been commercial cornucopia, but the accumulation (so far) of 75 tonnes of plutonium and 3336 tonnes of uranium, all stored and closely guarded but with no obvious use.

Not all the issues highlighted by the objectors have proved to be valid, but clearly some have and the resistance to THORP, although ineffective in the end, did create a powerful anti-nuclear movement which has had a significant influence on subsequent energy policy debates. Thus, when new reactors projects emerged, they were strongly opposed by increasingly well-organized objectors, some of whom could marshal extensive technical arguments – as occurred at public inquiries over the proposed PWRs at Sizewell (from 1983 to 1985) and at Hinkley (in 1988). As with THORP, it is true that in neither case were the objectors successful, but the tone of the debate had changed and the nuclear industry was much more on the defensive.

Opposition around the world

Similar developments occurred elsewhere around the Western world. The move to expand nuclear power following the 1974 oil crisis had led to increasingly militant grass-roots reactions. In the United States there were mass demonstrations at nuclear sites, for example, in May 1977 at the site of the proposed reactor at Seabrook in New Hampshire where 1400 people were arrested. These actions were non-violent, but in Europe larger and more militant demonstrations took place: for example, in November 1976, 30,000 people attended what was planned to be a peaceful demonstration against the planned reactor at Brockdorf in Germany. Some 3000 tried to occupy the site and there were violent clashes with the police who used water cannons, tear gas grenades and baton charges to try to restore order. Similar battles took place at Grohnde near Hamelin in March 1977. In July 1977, during a major demonstration against the French prototype Fast Breeder reactor at Malville, involving more than 60,000 people, one demonstrator was killed.

The demonstrations never the less continued: in September 1977, 60,000 people protested at the site of the proposed fast breeder at Kalkar in Germany, and, in the same month in Spain, 100,000 people joined a protest in Zaragossa, while 600,000 took part in a demonstration against plans for a reactor at Lemoniz (Elliott, 1978). One result of this level of agitation was the creation of powerful national and international networks, with, for example, the German anti-nuclear movement being particularly militant and influential.

Their influence was further strengthened by the nuclear accidents at Three Mile Island (in the USA) in 1979 and then more catastrophically at Chernobyl (in the former Soviet Union) in 1986, which provided what many saw as the final proof that nuclear power was not a safe option. Thus by the end of the century, around most of the world, the mood, amplified by the media, was thus increasingly anti-nuclear.

Why anti-nuke?

With Chernobyl still in the future, what was it about nuclear power that stirred up such great passions in the 1970s. In the 1960s, in the UK for example, conservationists had generally acknowledged nuclear as better for the environment than coal mining or dam building. Why the loss of support for an industry that claimed to be the shining hope of a brave new technological world, that was supported in the United Kingdom by all shades of political opinion from radical trade unionists to conservative businessmen, that was the ultimate in the 'progressive' dream?

The reasons for the emergence of the anti-nuclear power movement in the United Kingdom in the 1970s are obviously complex. Like all technological assessments made by the public, the reasons were only partly based on what scientists would call a 'rational' evaluation of the risks and benefits. There were of course many 'technical' concerns, but the movement was at heart an emotional response to 'nuclear fear', to dystopian images of an atomic future laid down since the beginning of the twentieth century. That the movement emerged in the early 1970s was in part due to opportunistic reasons – the public collapse in confidence in government institutions and authority, as well as internal dissent within the nuclear establishment. The nuclear proponents in the 1970s persisted in seeing opposition to nuclear power as simply 'emotional'.

The public expressed its anxieties over nuclear power through explicit questioning of reactor safety and the dangers of radioactivity, despite dismissals of such fears by inquiry inspectors. Furthermore the inability of the public authorities to provide a convincing answer to the frequently asked question (dating back at least to the first nuclear inquiry at Bradwell in 1956) of why, if nuclear power stations were so safe, were they built in remote areas? And why were there such detailed emergency measures in the event of any release of radioactivity? During the campaign at Stourport in Worcestshire, in 1970 against the proposed building of a nuclear power station, letter writers raised the issue of nuclear safety – one mentioned the Windscale accident of 1957. This theme of fear about nuclear safety dominated the campaign, despite reassurances from experts engaged by the county council.

Nuclear fear

The source and dynamic of the anti-nuclear opposition may have been emotional but they were not irrational. Spencer Weart in his book *Nuclear Fear* identifies four main themes that have influenced the way people thought about nuclear power (Weart, 1988, pp. 373–74). These were:

- The technical realities of reactors, both the economic opportunities and the hazards, as seen by scientists and transmitted to the public. From these realities particular 'facts' such as the hypothetical effects of low-level radiation in the event of an accident were selected and stressed.
- The social and political associations of nuclear energy, especially ideas involving modern civilization and authority, with nuclear reactors, became a symbol for all the evils of modern industrial society.

Nuclear power was singled out for this symbolic role largely as a result of

- the old myths about pollution, cosmic secrets, mad scientists and apocalypse that were historically associated with atomic power and radiation – indestructible myths with deep psychological resonances
- the threat of nuclear war, never for a moment forgotten.

Thus, to Schumacher, the old myths of scientists trespassing on forbidden territory were still valid. In an article 'Economics in a Buddhist country' written in 1955 he said that atomic energy was 'a prospect even more appalling than the Atomic or Hydrogen bomb. For here unregenerate man is entering a territory which, to all those who have eyes to see, bears the warning sign "Keep Out"'.

Reasons for opposition

What exactly were the campaigners against nuclear power opposed to? Their motives were diverse, ranging from NIMBY concerns through to opposition to capitalism. Motives for opposition to nuclear power can be divided into four categories:

- NIMBY – local opposition to any large-scale development nearby.
- Vested interest – opposition by the coal miners fearful of their jobs.
- Intellectual – based on aesthetic, ecological, ethical and economic reasons.
- Opportunistic – an opportunity by political groups to attack government policies.

NIMBY opposition is easy to understand and identify, and forms the bedrock of local opposition to proposals for nuclear reactors. In its early days it was termed the 'amenity issues' lobby – people opposed to developments in unspoilt countryside. Partly this was out of aesthetic concern by urban intellectuals for preservation of outstanding scenery, and partly due to the rural aristocracy and middle-class attempts to prevent economic developments that might undermine their privileges. For all large-scale developments in rural areas will bring benefits and costs to different sections of society, and nuclear power was no exception. Communities were often divided over the issue. The more vocal and well-organized middle classes, organized into ad hoc amenity societies, were however better able to put their views across and drown out less well-articulated working-class support. This occurred in 1956 over the Bradwell and Hunterston nuclear stations.

In the late 1970s, with the mushrooming of local groups opposed to proposed nuclear power stations, there were again accusations of NIMBYism. One sympathetic commentator wrote that 'the people who join such groups would probably oppose a coal-fired power station or an airport on the same site, and many are not perhaps, strictly anti-nuke' (Weightman, 1979, p. 311). Local middle-class opposition (in the early

1980s) to nuclear power stations was again accused of NIMBYism, portrayed by the media as 'piece-meal, irrational responses based in parochial concerns'. However such local groups were defended against these charges if they were willing to oppose nuclear power nationally and become part of a network of anti-nuclear organizations, such as the Anti-Nuclear Campaign or SCRAM. Thus while local nuclear opposition can be seen as NIMBY, it can redeem itself (in the eyes of sympathetic commentators) if it is willing to become part of a national network opposing nuclear power. To cynics this is simply 'greenwash' – the adoption of an environmental position to further self-interest.

Vested interest

The coal miners had a clear economic or vested interest in opposing nuclear power, due to their determination to stop job losses in their industry. The opposition of the coal miners, their union (the NUM), their supporters in the Labour party and the NCB to the expansion of nuclear power at the cost of coal in the mid-1960s has been overlooked by most historians of the anti-nuclear movement. As noted earlier, the campaign led by Alf Robens and Fritz Schumacher ended in defeat in 1968. Opposition to nuclear power from sections of the NUM however continued in the 1970s with Arthur Scargill, its president, involved in the creation of Energy 2000, which was part of the opposition at the Windscale Inquiry in 1977.

Intellectual dissent

There was criticism of nuclear policies and siting based on aesthetic, ecological, ethical and economic reasons but until the mid-1970s these were largely confined to academic journals and small-movement publications. Dissent initially came from industry insiders, concerned about the feasibility of too rapid an expansion of the industry, followed by criticism from academics critical of nuclear economics and the AGR reactor choice. There was little public discussion of radiation or safety issues, until Schumacher raised the issue in a speech in 1967, which proved highly controversial but was quickly forgotten. Few people saw the linkages between nuclear weapons and nuclear power, and while the public were fully aware of, and campaigned against, the dangers from radiation fallout from nuclear testing, they appeared unaware of radiation emissions from nuclear reactors.

This changed in the early 1970s due to the books from the United States which publicized the safety hazards and dangers of radiation emissions from nuclear reactors. American campaigners quickly adopted these arguments against nuclear reactors, after initially relying on ecological ones about 'thermal pollution', disturbance of habitat and aesthetic damage to the landscape. The concerns of Sternglass over the hazards of low-level radiation were given wide publicity by the national press in Britain. Very rapid scientific dissent over 'safe' limits for radiation exposure, and also the unresolved problems over radioactive waste storage, became translated by eco-activists into the prime reason for the public to oppose nuclear power.

Opportunistic dissent

Intellectual critics of nuclear power were also concerned with issues of democracy and equality, seeing nuclear institutions as examples of remote and overbearing bureaucracies that threatened civil liberties and must therefore be curbed. This criticism of the 'nuclear state' is built on previous criticism of the modern technocratic state and the power of corporations, such as by Theodore Roszak and Lewis Mumford. Nuclear power in the United States thus became a rallying point for student radicals and other social critics, particularly after the end of the 'campus wars' and demonstrations against the Vietnam War in the early 1970s. Previously the new (and the old) left had shown little interest in environmental or nuclear power issues, sometimes attacking it as a distraction from more serious social issues. In the United States protest against nuclear power (like environmentalism) was seen as a middle-class provincial movement concerned with NIMBY issues. Whereas in Europe it was seen as a means of uniting peasant farmers with students and as one of the most effective rallying points around which European social critics could gather. This can be seen in Britain by the gradual adoption of the anti-nuclear power cause by the new left, starting with the Socialist Workers Party in 1976. It also attracted religious, peace and women's groups who had a long-established concern with nuclear weapons.

The end of the nuclear dream

The fall from grace of nuclear power had more to do with the perceived shortcoming of the nuclear bureaucracy than with technical failings. In the United States, the AEC, and in Britain, the AEA, were most probably no worse than any other bureaucracy, it was just that the public had such high expectations of the nuclear endeavour after decades of propaganda. When the nuclear utopianism collided with the grim realities of the 1973 'energy crisis', of power shortages and petrol queues, disillusionment rapidly set in. This time the factors that had restored confidence in the mid-1950s were under attack, namely public trust in business and government leaders and faith in the ability of science and technology to solve social and economic problems. The problems with nuclear power were as always, but now the public no longer had faith in the ability of government, business and science to solve them. Worse was to come, with the Three Mile Island and Chernobyl disasters, which to many people confirmed the view that not only was the technology dangerous, but the nuclear industry and associated bureaucracies could not be trusted.

So what of the dreams of the atomic age bringing unimagined benefits for most of humanity? Hilgartner in his book *Nukespeak* lays the blame for the failure of nuclear power on the 'nuclear mindset' of its promoters, arguing that

> Time and time again, nuclear developers have confused their hopes with reality, publicly presented their expectations and assumptions as facts, covered up damaging information, harassed and fired scientists who disagreed with established policy, refused to recognize the existence of problems, called their critics mentally ill, generated false or misleading statistics to bolster their assertions, failed to learn from their mistakes, and claimed that there was no choice but to follow their policies. (Hilgartner et al., 1982, p. xiv)

A more measured response is from Steven Cohn who sees the tragedy behind an idealistic venture, and he ended his book *Too Cheap to Meter* with the words:

> The sad conclusion from my perspective is that the nuclear dream has not worked out. The technology has failed and should be put aside until other energy options have been exhausted and the industrial subculture that nurtured the first nuclear era dismantled. I find this a sad conclusion because the nuclear dream was compelling, the imaginations behind it were talented, and the human energy and economic wealth mobilized to pursue it were enormous. (Cohn, 1997, p. 318)

While environmentalists may have failed to stop THORP and other nuclear projects, they have hopefully laid the grounds for better decision-making and public participation in nuclear power issues (Pearce, 1979). Whether that will be reflected in the outcome of the current debate over climate change, and the proposed nuclear revival, remains to be seen.

References

Cohn, S. (1997) *Too Cheap to Meter: an economic and philosophical analysis of the nuclear dream*. State University of New York Press, Albany, NY.

Elliott, D. (1978) *The Politics of Nuclear Power*. Pluto Press, London.

Herring, H. (2005) *From Energy Dreams to Nuclear Nightmares: lessons for the 21st century from a previous nuclear era*. Jon Carpenter, Charlbury, Oxon.

Hilgartner, S., R. Bell and R. O'Connor (1982) *Nukespeak: nuclear language, visions and mindset*. Sierra Club Books, San Francisco.

Pearce, D., L. Edwards and G. Beuret (1979) *Decision Making for Energy Futures: a case study of the Windscale Inquiry*. Macmillan, London.

Weart, S. (1988) *Nuclear Fear: a history of images*. Harvard University Press, Cambridge, MA.

Weightman, G. (1979) 'The anti-nuke explosion', *New Society,* 8 November, pp. 310–311.

Further reading

Aubrey, Crispin. *Meltdown: the collapse of the nuclear dream* (London: Collins & Brown, 1991).

Bolter, Harold. *Inside Sellafield* (London: Quartet Books, 1996).

Boyer, Paul. *By the Bomb's Early Light: American thought and culture at the dawn of the atomic age* (New York: Pantheon Press, 1985).

Breach, Ian. *Windscale Fallout: a primer for the age of nuclear controversy* (Harmondsworth: Penguin, 1978).

Bunyard, Peter. *Nuclear Britain* (London: New English Library, 1981).

Ford, Daniel. *The Cult of the Atom* (New York: Simon & Schuster, 1982).

Hall, Tony. *Nuclear Politics: the history of nuclear power in Britain* (Harmondsworth: Penguin, 1986).

Patterson, Walt. *Going Critical: an unofficial history of British nuclear power* (London: Paladin, 1985).

Roberts, Fred. *60 Years of Nuclear History* (Charlbury, Oxon: Jon Carpenter, 1999).

Walker, William. *Nuclear Entrapment: THORP and the politics of commitment* (London: IPPR, 1999).

Welsh, Ian. *Mobilizing Modernity: the nuclear moment* (London: Routledge, 2000).

Williams, Roger. *The Nuclear Power Decisions: British policies 1953–1978* (London: Croom Helm, 1980).

Wynne, Brian. *Rationality and Ritual: the Windscale Inquiry and nuclear decisions in Britain* (London: British Society for History of Science, 1982).

Part II Structuring the Debate

4
Building or Burning the Bridges to a Sustainable Energy Policy

Gregg Butler and Grace McGlynn

For those interested in the areas of energy and environmental policy within the United Kingdom, there has been a bit of a 'phoney war' going on over recent years. The energy review of 2003 left the nuclear energy option open but said 'not yet'. Then the government said that there would be a decision about whether to go down the 'new build' route within the current parliament. At the Labour Party conference in Brighton in September 2005, the prime minister mentioned the 'n word' as a potential means of ensuring security of energy supplies at a time when oil prices have risen, China has become the third largest consumer of fossil fuels and global warming is being blamed for everything from Hurricane Katrina to a fear of the demise of champagne production in France. And then amidst great excitement and angst a UK Energy Review was announced – and something depressingly familiar happened.

The 'nuclear debate' as in all its previous rounds, is being dominated by the extremes. The industry, to whom the benefits of nuclear often seem to be a near-religious belief, and the Greens, who see total fear and an absence of benefit, fight it out in the media. Faced with the answers of 'absolutely yes' and 'absolutely no', small wonder the public is unlikely to be able to judge what the 'ground in the middle' of the extremes look like. And for the media, it's a gift. Greenpeace and the industry slugging it out on the Today Programme without a mediating voice in sight … 'that was a nice dust-up … and now Gary Richardson with the sport …'.

The Greens disapprove of nuclear for various reasons, starting with the connection between nuclear power and nuclear weapons, a conviction of the severity of possible accidents and the dangers of nuclear waste, all against the background of a general dislike of big organisations and a centralised society, which are seen as necessary for the deployment of nuclear power. In the United Kingdom, these concerns are particularly focussed by the Magnox power stations, for which reprocessing, with the separation of plutonium, is a direct prerequisite of their continued electricity generation.

But surely global warming and energy security are rather more serious than this and demand a rather more searching debate. If global warming is indeed a fact, and if the human consumption of fossil fuel is a serious contributor to climate change, then the changes that will come if we do not act to moderate our behaviour will be globally serious and will threaten 'civilisation as we know it'. If global warming is real, we must seek a more constructive way to define public policy than leaving both the public and their political representatives to choose between two extreme views – trying to find the middle ground between 'yes' and 'no'.

So how to get big topics discussed in a meaningful way? One way is to see what areas can be agreed on, what 'facts' are agreed by all parties, and where differing value sets either change perceptions of the 'facts' or prevent agreement on what the 'facts' are. The value of even limited agreement has been well summed up in the nuclear context:

> When former opponents find common ground, their joint recommendations carry far greater weight than they could ever hope to achieve alone. The whole from a stakeholder dialogue is much greater than the sum of its parts
> Michael Meacher MP, formerly UK Environment Minister, 12/4/00

The authors will now 'come clean'. We have both worked in, or for, the UK nuclear industry for many years. We have both gone through the phase of being evangelical, but 6 years participation in the BNFL National Stakeholder Dialogue process (BNFL, 2006) has replaced certainty with a willingness to see other viewpoints. We do, however, believe in science, and believe that as many facts – agreed and verifiable facts – should be gathered as a background before any decision-making process starts, and that such a process is based on proactive and collaborative stakeholder engagement.

Stakeholder dialogue

Most of us who have been involved in stakeholder dialogue, whatever 'camp' we started in, recognise several phases in helping decision-making:

- Agree the scenarios to be examined
- Agree the attributes of an acceptable solution – and which direction 'good' is

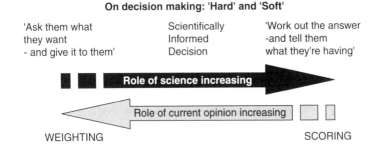

Figure 4.1   Decision-making approaches

- Agree the 'facts' – and joint fact-finding is often crucial here
- Undertake a process to define the pros and cons of each scenario – and if consensus cannot be reached – find out why not

In fact, some element of 'joint fact-finding' – groups containing people with very differing viewpoints seeking areas where they can agree and turn 'facts' into *facts* – is often a prerequisite for agreeing on scenarios. If, however, the staged process can be taken to its conclusion, then you will have a clear picture of the agreed elements of any decision-making field and the areas of, and reasons for, disagreement. This will then provide the problem holder with much of the background needed to make a decision. This will inevitably be on the continuum illustrated in Figure 4.1.

Consensus?

In any controversial area, full consensus will not be achieved. There will be those on both sides whose beliefs will not admit the consideration of any version of the 'truth' but their own. However, if a large majority can agree on a tranche of basic facts, then the arguments can be deployed on the basis of real differences based on really different value sets. All parties may not be able to *share* these honestly held views, but at least they can be respected, and mutual respect is the only possible background to achieving a result that 'we all can live with'.

Sieving

'Sieving' means 'ruling out options or scenarios on the basis of a single parameter being unacceptable' – the option won't pass through the sieve and is discarded. If a variety of groups with a variety of values and opinions apply sieving, especially without an agreed fact base, then the chances are that you'll have nothing left!

For example:

- Onshore wind is noisy, kills birds, and ruins the landscape – out!
- Offshore wind is expensive, has a low load factor and just won't do the job – out!
- Nuclear is dangerous and has a dreadful waste problem – out!
- Biomass – in the UK's climate – you must be joking! – out!

and so on. Sieving in the absence of agreed 'facts' is a recipe for polarisation and the mortal enemy of meaningful debate.

Global warming

If we are convinced that global warming is a real problem then surely

- The time for change is now, or more accurately 'was then'
- We believe that the ship is in danger of sinking
- And if we really believe this, we'd be looking for the pumps, any pumps, to save us ... not balancing angels on pins and disagreeing on detail!
- Can we actually do anything – in a situation where the hint of a fuel protest has half the country queuing at garages all day?

This is surely the nub of the energy conundrum: all the prognostications of what to do involve sacrifices – either of aspiration or actuality. All of the future scenarios, when viewed from democracies, involve some element of 'Turkeys voting for Christmas'. All of the future scenarios, when viewed by politicians – have disquieting elements of being unable to please all the people all of the time.

There has to be scope for agreement by a broad coalition of interests on what the options are from a scientific point of view. This is important, because while it is sometimes very nice to hope that 2 + 4 will equal 5, it very rarely turns out to be true. The work of these studies must not be the work of science in isolation, but must be shared with the humanities and economics fields from the start. Such 'across the board' studies might usefully examine some basic aspects of energy policy, which might include:

1. Can we agree that electricity generation from renewables alone is intermittent? That renewables alone cannot 'solve' the carbon

emissions problem without non-credible assumptions on either installed capacity, usage patterns, or both? And that all credible short–medium term renewables scenarios need backup from other generation sources?

2. Can we agree that current nuclear designs are primarily baseload producers and that here too backup is required?

3. Can we agree that hydrogen is the only credible medium-term solution for the CO_2 produced by transport, and that 'how we generate the hydrogen' is a key factor to be examined?

4. Can we gain broad agreement on the rate of change in energy usage and/or cost which the UK population might accept as being in their long-term interest?

These questions and a suite of others like them need to be looked to inform a range of credible scenarios for the future. They need to be looked at by academics from a wide range of disciplines and they need to be simultaneously studied by a wide range of stakeholders. They need to be looked at in a setting where the study result is not preordained by the organisations organising it.

This approach may lead to some extreme interests challenging their current views. Views of the importance of global warming have already gained supporters for nuclear power, with James Lovelock and the late Bishop Montefiore being high-profile examples. Would more Greens 'put global warming first' if it was found that the most viable scenarios to combat climate change involved nuclear power? Could UK Greens become more amenable to nuclear power once the reprocessing connection is removed as Magnox and oxide fuel reprocessing cease?

It is not as if lots of work on scenarios and facts is not taking place. In just one example, the work of the Tyndall Centre (Tyndall, 2005) has produced scenarios giving several ways of 'beating global warming'. Some have nuclear some do not, but many of them demand real change in the way we do things as individuals and as a nation. These views and many others like them are communicated to the government. A proposal for changes as significant as those inferred by some scenarios will be controversial – and with the current protagonist/antagonist media culture it is easy to see why there is great political reluctance to launch a really inclusive public debate.

The position is not helped by the absence of a broad-based non-partisan advisory group on energy matters. True, the Sustainable Development Commission (SDC, 2006) has opinions and has made them known, but can hardly be described as objective in the area of nuclear power. When sixteen people all say either 'not now' or 'not ever' on a subject as keenly debated and divisive as nuclear power, they are either unrepresentative of the population or possessed of insights denied to that population. An examination of the Commission's report and its supporting papers show no new facts. Referring back to the diagram above we have a body of work which is largely opinion-led rather than fact-led.

This paper is not, however, about seeking to 'convert' those of extreme and set views. It is about whether or not the United Kingdom can face up to its energy future with a debate which centres round the possible rather than the fanciful, and attempts to involve and empower the middle ground to arrive at a future which makes scientific, economic and social sense. In the interests of a sustainable energy future for the United Kingdom, it must be hoped that this can be done – but the portents are not good.

References

BNFL (2006) *National Stakeholder Dialogue*, British Nuclear Fuels Ltd: www.the-environment-council.org.uk.

SDC (2006) *The role of nuclear power in a low carbon economy*, Sustainable Development Commission, London: www.sd-commission.org.uk.

Tyndall (2005) *Decarbonising the UK*, Tyndall Centre for Climate Change: www.tyndall.ac.uk.

5

Criteria for a Sustainable Energy Future

Paul Allen

Our future energy strategy, in fact our very survival hangs on three key challenges. Firstly, our leading climate scientists now conclude that if global greenhouse gas emissions exceed the planet's critical 'tipping point', it will set us on course for abrupt, accelerated or runaway climate change. Based on an extensive review of all the relevant scientific literature, many leading climate scientists now conclude that we have a decade before we cross a crucial 'tipping point' where average global surface temperature rises to more than 2°C above its pre-industrial level.

Exceeding the planet's 'tipping point' sets us on a future course for abrupt, accelerated or runaway climate change. Runaway feedback in climate change could entail massive agricultural losses, widespread economic collapse, international water shortages, massive rises in sea levels, a shutdown of the Gulf Stream, refugee problems on a scale not yet experienced – a complex of global catastrophes on a scale that would dwarf the recent events in New Orleans and run for tens of thousands of years.

Based on our current knowledge, if we are to keep the average global surface temperature below this crucial tipping point, we must keep the atmospheric concentration of the greenhouse gas, carbon dioxide, below 400 parts per million (ppm). Before we began to industrialise, CO_2 levels were around 280 ppm, and today they have reached around 370 ppm and are still rising. So at the current rate of rise we will cross the tipping point in around decades. Across the developed West we must rapidly break our addiction to cheap fossil fuels in order to reduce our CO_2 emissions by 60–80% to remove the cause, while also protecting and expanding the capacity of the world's forests and soils to respond to the problem by drawing down the excess CO_2 from the atmosphere.

Secondly, our unstoppable thirst for oil is being halted by the immovable facts of geology. Rather than talking about when oil could 'run out', the peak oil experts predict that despite accelerating demand, the global rate of production may be at, or approaching, its peak. The Association for the Study of Peak Oil and Gas (ASPO) is a collection of industry figures, politicians and academics. From quiet beginnings three years ago, ASPO is no longer 'at the fringes', it is now being taken very seriously in many quarters.

The world is using more oil than it finds, and discoveries of oil fields peaked in the 1960s. Despite technological advances since then, new field discoveries are at an all-time low. This has led to the current lack of any 'cushion' between supply and demand, and to the consequent high prices. The outcome for the world, if ASPO is correct, is catastrophic (ASPO, 2006).

The economic growth in India, Asia and China has exceeded all industry speculation. Car sales in China are expected to total up to 10 million vehicles annually by 2010. In Beijing alone, more than 1000 new cars hit the city's streets each day. Many analysts now suspect the petroleum joy ride of cheap, abundant oil which has sent the global economy whizzing along for decades may be coming to an end. Some observers of the oil industry predict that this year, maybe next – almost certainly by the end of the decade – the world's oil production, having grown exuberantly for more than a century, will peak and begin to decline. And then it really will be all downhill. There will be warning signs: prices will rise dramatically and become increasingly volatile; with little or no excess production capacity, any supply disruptions such as hurricanes in the Gulf of Mexico will drive world oil markets into frenzy; as will occasional admissions by oil companies and oil-rich nations that they have been overestimating their reserves. Does any of this sound familiar?

Our third challenge is one of global equity. The developed countries use most of the energy, and everything else for that matter. Bangladesh, for example, uses less than 200 kg per head of oil equivalent per annum, compared with nearly 4000 kg per head in the United Kingdom. Net CO_2 emissions per head in the United Kingdom are 50 to 100 times more than those of the people in Bangladesh or Tanzania. Despite record increases in global economic activities, the rich are still getting richer and the poorest are being left behind. So-called third world development is not working. The majority of the world is expected to pull itself out of post-colonial poverty through globalised trade, so we can satisfy our selfish desires for cheap exotic products. Firstly, the sheer scale will be more than the climate can bear and secondly, there isn't enough cheap oil left to make it a long-term solution. Business as usual simply does not work from the climate's point of view; neither does it work from an energy reserves point of view, and it certainly does not work from the point of view of global equity.

Addiction

Although these challenges are becoming increasingly familiar individually, their respective experts still work in relative isolation and their solutions are rarely considered in unison. There are solutions to peak oil which accelerate climate change, and there are solutions to global equity which exacerbate peak oil. These kinds of measures – solving one challenge at the expense of another – will not solve the problem. The key to our future survival is to solve the three main challenges together, and do it in a way which also encompasses our personal well-being. Once we join the dots and look for the bigger picture, we will find plenty of common ground. In fact, facing up to our oil addiction and re-thinking our diet, buildings, energy, water, work, clothing, heating, holidays and healthcare can mitigate climate change, help protect us peak oil while releasing resources the majority world urgently needs.

We see the crisis, we see the solutions – but our almost total failure to act makes it increasingly obvious that our entire culture, indeed our entire civilisation, is locked into 'fossil fuel denial'. Denial is the primary psychological symptom of addiction. It is both automatic and unconscious. Addicts are often the last to recognise their disease, pursuing their addictions to the gates of insanity as their world collapses around them. Denial defends the individual or collective consciousness from some truth which they cannot afford to acknowledge because it would expose overwhelming feelings of fear, shame or confusion. As long as we remain in denial about climate change, peak oil or the suffering of the majority world we are free from any painful feelings, and can lose ourselves in our affluence.

Denial is not the only symptom of addiction. When the supply of the addicted substance is restricted or becomes uncertain, it is common for anti-social behaviour to emerge on a collective or individual level. Nothing is more important than the addiction itself. Everything is geared towards getting the dependence met, and the deeper into addiction we go, the greater the selfishness. In the case of powerful addictions, the addict will even break national or international law in order to secure their supply. Consequently, addicts often hide their behaviours from others or create smokescreen excuses to justify their anti-social behaviour. At long last, following this year's State of the Union Address, America and her over-developed allies now admit they are addicted to oil. The recovery plan must not stop here however, the next step is to re-think how we use energy to deliver our well-being to reduce the usage rate and so curb the anti-social behaviour.

Why is a fossil fuel addiction so hard to break?

Fossil fuels are a hard act to follow. Around 400 million years ago, the Earth's atmosphere was rich in carbon dioxide. A verdant carpet of plants covered the land, soaking up the sun, as did the surrounding oceans of single-celled algae. This process went on for millions upon millions of years. The vast surface area of the Earth was turning the sun's energy into plant and animal matter, concentrating it into carbon-rich deposits of coal and oil and gas.

This 70 million-year reserve of ancient sunlight lay for further eons until the industrial revolution when the United Kingdom and then others started using coal on a massive scale and then, in 1859, oil was discovered in Pennsylvania, USA. This represents a critical moment in human history, for up to that point most of humanity had been fed only by the annual sunlight falling on croplands. With the advent of first coal, then oil and gas humanity started living off our planet's 'sunlight-reserves' on an ever-increasing scale. We had discovered a massive energy 'bank account'. As a result of this abundant cheap fossil-fuel bonanza, we based our global models of industry, commerce, food production, finance and habitation around them.

Consider a typical day: we wake up in a house built by people we do not know, with materials from we know not where. It is heated by fossil fuels manufactured and brought to us in ways we are only dimly aware of, by an energy utility now owned in America. The house is located in an area where we know few neighbours and is mostly owned by a German bank. We breakfast on food grown heaven knows where, by unknown hands, using methods we never see and cooked we know not how. Our lunch is bought from shops that would be empty in three days without fuel. Our waste travels to unknown destinations, to be treated in ways we could not understand, by people we will never meet. We pay for it all through a Japanese bank over which we have no control. We now depend for our continued existence on increasingly remote suppliers through ever-expanding systems that have no obligations to us, and indeed are not expected to have any.

Oil and gas reserves were built up over hundreds of millions of years, and we have used half of them in only 150 years. They are the most concentrated, transportable, convenient fuel we have ever had – or ever will have! Recent events, at home and abroad, force us to question globalisation and the increasing influence of trans-national corporations. The changes it brings are not made for the benefit of farmers or consumers; they are made to increase the profitability of the globalised distributors. We must now question which aspects of our lives can we still trust to multi-national supply chains and which aspects are better sourced locally. By the way we have developed our economy, industry, food production, transport systems and habitation; we have created an addiction to a continual supply of abundant cheap petrochemicals.

So what do we do?

The global media routinely carries – 'the end of the world' articles – which journalists might think are raising awareness of climate change. The reality is that doom-laden coverage makes readers become apathetic about the threat. Recent research, by the green communications agency Futerra, found that 60 per cent of articles about climate change in national newspapers were negative and failed to mention possible solutions. Only a quarter included any mention of what could be, or is being, done to fight climate change. Warnings without any sight of a solution may result in denial, depression or even trigger one last hedonistic binge. Worse still, we may panic, disagree badly about who has the right solution, and fall into such an entrenched fundamentalist standoff, so nothing gets done at all.

Unfortunately this seems to be exactly what is happening between advocates of nuclear and renewable energies. Over the past few decades, a couple of major accidents plus the accumulating problems of both waste and costs have driven the nuclear industry far from public grace. However nuclear advocates are now jumping on the climate change bandwagon in hope of a last-minute reprise. In the other corner, the renewables industries, having been very much in poor relation to nuclear for several decades, are of course now reluctant to lose their long awaited chance for the limelight and some realistic research and development funding. So the debate whether to allocate vital, but finite energy research and development funding to either nuclear or renewables is an important one.

However, neither new nuclear nor renewables have any hope of making an impact on the scale required unless we rapidly make a concerted effort to implement an energy descent plan. Too many years of too cheap energy has led us into some embarrassingly wasteful practices. Many of our power stations give out as electricity only one-third of the energy we put in, the remaining two-thirds going up the cooling towers rather than heating nearby homes. And electricity is not the only problem; we must also get to grips with our profligate use of solid and liquid fuels. There are now so many people making single passenger car trips that the roads are literally clogging up. Subsidised cheap air flights are escalating almost exponentially and we are still building inefficient, badly designed homes in thousands.

Powerdown: creating a carbon descent pathway

We are energy obese; we use far more energy than is good for us. Richard Heinberg's original concept of 'Powerdown' is a sane response to humanity's increasingly grave situation, and deserves further exploration (Heinberg, 2004). As we powerdown our energy requirements, delivering them with renewable sources not only becomes achievable, it rapidly becomes cost competitive as oil prices soar, and significantly more reliable as fossil energy supplies falter.

Above the basic level needed to provide food, clothes and shelter, using extra energy does not necessarily make us any happier. Since the 1970s the UK's GDP has doubled, but our perceived 'satisfaction with life' has hardly changed. We've become energy obese, using far more energy than is actually required to deliver our well-being.

Powerdown is not the same as energy efficiency: it goes very much further. We drive an oil-powered machine to plough the land, and another to plant the seed. We then use fertilisers and pesticides made with oil, and irrigate with water pumped by oil. We harvest the crop with oil-powered tractors and process it with fossil-fuelled electrical equipment. Finally it is packed in plastic and driven further than you ever imagined. The bottom line is that we eat ten calories of fossil fuel energy for every calorie of food we consume.

The potential powerdown which could be achieved through a re-think of our food alone is massive. For example, we export many thousands of tonnes of lamb to the EU, while also importing many thousands of tonnes from the EU. Similar paradoxes exist for most other products, this is outlined in the recent 'UK Interdependence report' from the New Economics Foundation (NEF, 2006). Switching to a locally sourced, mostly organic, less-processed, low-meat diet will not only increase our general health and well-being, it can massively reduce the fossil fuel dependence of our eating habits and create systems which are considerably more reliable.

But could we really feed ourselves if we broke our addiction to oil? Firstly, it would require a change of diet, but that's something we need to do in any case. Secondly, supermarkets reject around 30% of vegetables because they are the wrong shape, colour or size. Further waste occurs when food is processed into ready meals. Finally, consumers bin about 30% of what they buy. If we stopped this wastage then we could be far more self-reliant in food and vastly reduce the oil needed to provide it.

The simple fact remains that a powerdown approach is absolutely essential and must form the cornerstone of any proposed way forward. Initially saving energy will save us money, and even after that, it is considerably cheaper than generating more energy, particularly if we now must include all the externalised costs such as environmental degradation.

Although capturing and sequestering the carbon released from burning fossil fuels does appear to offer a potential solution, it is very early days, and it would be risky to predict its role as a form of low carbon energy generation. However, it is clearly an area well worth serious investigation and research, as it may give us some nice surprises in the future. So as we are all agreed that rational use of energy is both absolutely essential to give a realistic target for new low carbon generation technologies, and the best value place to start, implementation must be accelerated immediately. Our current energy obesity is clearly unsustainable, once we powerdown it gives a more realistic target which can be met by alternative forms of generation.

Breaking the deadlock: a low carbon generation technology

The next key challenge is to break through the deadlock between nuclear and renewables so we can decide which will be the required low carbon energy generation technology. To do this we must follow the approach any engineer worth their salt would take to select a technology for a particular application. First they would meet with clients, customers and other key stakeholders to agree the 'selection criteria', and then evaluate all available options against these agreed specifications. In my mind, for the nuclear/renewables debate the selection criteria would include:

- What is the environmental impact?
- Is it technically feasible on the scale required?
- What is the impact of 'trend setting' overseas?
- What is the full life cycle cost?
- Does it make us vulnerable to terrorist attack?
- Is the speed of installation fast enough?
- What about security of supply?
- Does it create jobs and exports?

- What impact does it have on civil liberties?
- Can it access heat & liquid fuel markets?

Clearly both sides will have some passes and some fails; however, when we see the results all together, some interesting patterns are revealed. Firstly, to avoid talking at cross purposes about what we mean by environmental impact, it is vital we distinguish between the 'global habitat', 'human health' and 'perceived visual/aesthetic' environments, so their respective impacts can be weighted appropriately. The impact of fossil fuels in the 'global commons' environment is very different from the impact of a nuclear meltdown on the human health environment, which is clearly different from the impact of wind power in the perceived aesthetic environment.

It is also vital to compare like with like – we must always look at the full life cycle costs. Nuclear fuels have got huge environmental and social costs that are only now coming up for payment. UK nuclear decommissioning and clean up liabilities from the past 30 years have been estimated to be around 56 billion pounds. It is just too easy to 'discount' these costs into the future; the full cost of plant decommissioning and safe transport and storage of waste materials must be taken into account in the cost of electricity paid for by the customer. If we then add the cost of the research and development, the military and civil cost of protecting each gram of fissile material, plus the ecological costs and the hidden subsidies, the true cost of new nuclear looks truly prohibitive. One would of course do the same exercise for the renewables, costing the whole life cycle including decommissioning and clean-up.

How did each technology score?

Nuclear energy: the passes

- Technically feasible for base load electricity generation
- Low visual impact
- Would create UK jobs
- Low relative health risk during normal routine operation (in the short term)

Nuclear energy: the fails

- Worst-case accidents are very serious and very costly
- Long-term effects of low level 'routine emissions' are uncertain

- Problem of nuclear waste not solved after 40 years
- Full life cycle cost (including realistic decommissioning and wa storage costs) is prohibitive and heavily 'back end loaded' making it unpredictable
- Poor financial liabilities of British energy for decommissioning existing nuclear plant
- Unattractive to free market investment
- Nuclear facilities are attractive targets for terrorism and 'recreational malice', amplifying the effect many times due to long-term effects of the fallout
- Not globally replicable – nuclear power must be limited to 'safe states' without any potential for political collapse or misuse of the technology
- No agreed international mechanism for this control (e.g. USA/Iran)
- If other nations do follow nuclear trends, large amounts of nuclear materials will be in constant global circulation (dirty bombs, etc.)
- Speed (public acceptance and track record of delays in construction)
- Uranium reserves are finite
- Civil liberties (restrictive legislation required)
- Unattractive as domestic CHP (Combined Heat and Power)
- Not currently available as liquid fuel

Renewables: the passes

- Low environmental impact in most categories
- Low risk of catastrophic failure (except dams)
- Technically feasible – the energy is out there
- Secure against fuel running out
- Secure against fuel price rises
- Secure against 'political' fuel blockades
- Average seasonal generation is predictable
- Tidal energy is very predictable
- Unattractive to terrorists and 'hackers'
- 'Full life cycle costs' are increasingly competitive
- Attractive to free market investment
- Labour intensive – lots of UK jobs for steel, shipyards, post north sea technologies, engineering and electronics
- Massive export potentials – no problems in encouraging other countries to adopt renewable technologies

- Low impact on our civil liberties
- CHP feasible with biomass

Renewables: the fails

- Sometimes, but not always, have a visual impact
- Speed of future installation depends on effective R&D investment strategy now
- Not currently widely available as liquid fuel (although bio-diesel is increasing fast)
- Intermittency of some renewable energy sources causes some extra costs and modifications of the grid

Analysis

Firstly, we must also recognise that the so-called war on terror has made the world a very different place and so our policy for dealing with climate change, peak oil and related energy strategies must reflect this change. Nuclear generators, enrichment, storage and reprocessing plants make the UK extremely vulnerable to the consequences of a terrorist attack. Imagine what even a modest attack on Sellafield could do to the economy, tourism (and the population) of the majority of northern England. By comparison, renewables are a diffuse energy supply and so, with the exception of large hydro projects, it would take a substantial amount of bombs to make any noticeable impact on our energy supply.

Perhaps the key aspect to breaking the nuclear/renewables stand-off is in recognising the need for global-scale co-operation, and the Global Commons Institute's 'Contraction and Convergence' model is leading the field in this respect (Meyer, 2000). Even if the UK met all our climate change targets by the end of the year, humanity can only avoid crossing the climate change 'tipping point' if other countries meet their targets too. If the United Kingdom makes nuclear power, the core component of its response to climate change, many other rapidly developing economies will want to follow suit. It will then be very hard for the 'developed nations' to make a case why we are allowed nuclear technology, when the countries which manufacture our fridges, cars and cookers are considered too 'unstable' to be granted access to the same technology, yet must also meet their climate change targets. The US attitude to uranium enrichment in Iran is a clear example of this. Under the Nuclear Non-Proliferation

Treaty, a country is allowed, under inspection by the IAEA, to enrich uranium to a level needed for civil nuclear power. The United States and some Western countries say that Iran should not be allowed to develop enrichment technology at all because it cannot be trusted. Is nuclear power really an issue we want to fall out over and can our global response to climate change and peak oil really be limited to only the countries which United States trusts? The UK Sustainable Development Commission, in their report *The Role of Nuclear Power in a Low Carbon Economy* reaches the following conclusion: 'If the UK cannot meet its climate change commitments without nuclear power, then under the terms of the Framework Convention on Climate Change, we cannot deny others the same technology' (SDC, 2006).

Finally, as we try to reduce our addiction to fossil fuels we should also be very cautious of swapping one addiction for another. Some technologies are very hard to stop once you pick up the habit. Britain must maintain expensive nuclear facilities far into the future, if it is to safely deal with its waste and decommissioning liabilities. Indeed, if the Romans had used nuclear power, we would still need to be tending their wastes, which I am sure we would be happy to do in their honour, but what would have happened to it during the Dark Ages? If we take another nuclear fix, many other countries will want a try. If we want to stem the flow, we should quickly re-brand the United Kingdom as 'Beyond Plutonium'.

Renewables

It has become clear that if our energy strategy is based on exploiting ever-diminishing reserves of nuclear or fossil fuels, the inevitable demand-driven improvements in extraction technology may increase short-term yield, but only at the cost of depleting the reserves even faster – thereby making the problem worse! But if we switch from energy reserves to energy flows, the same demand-driven improvements in extraction technology will increase annual yield, but on a permanent basis.

Renewables can work on both a national, regional and domestic scale. We know the energy is out there. Britain has all the skills required and the best renewable resources in Europe. We have an offshore energy industry in the North Sea which has peaked and is now ripe for conversion to marine renewables. We can use the sun to heat water, buildings and to make electricity. If only one-third of current electricity consumers (approx 10 million) installed microgeneration systems such as wind, hydro or solar electric at around 2 kW, generation levels would be similar to that from current nuclear capacity.

A 'contraction and convergence' to our global fair share of fossil fuels does not mean a return to pre-industrial energy levels. A wide range of renewable energy generators can vastly increase the amount of energy we can capture from carbon free flows which are constantly replenished. It is fairly easy to predict the amount of energy we can capture over a season; we just can't tell exactly which day we will get it. But, provided we have a good spread of technologies and a good spread geographically, the problems of intermittent supply can be overcome.

Conclusions

Those who are old enough to remember liken our current situation to how they felt in 1939 Britain. We know something big is just over the horizon, we know that it will be a harsh challenge, and we are not sure how society is going to cope. Yet in 1939 when push came to shove, the various factions pulled together offering a single united response. This is what we need most urgently now, but on a global rather than national level. A shift of energy policy from the current inefficient use of fossil and nuclear fuels to energy efficiency and renewable energy sources is not only imperative for keeping us below the tipping point, it is also vital to our international security.

In every crisis there is opportunity. It is vital that we make it clear that this transition does not have to be that painful. To succeed, we must remain optimistic and focus on the benefits. If humanity can act in unison as a global community, a positive outcome is possible. Although it is an unprecedented challenge, there are encouraging examples. With the loss of Soviet oil in 1990, Cuba was forced to undergo an astounding transition from large plantations reliant on fossil-fuel-based pesticides and fertilizers, to small organic farms and urban gardens. Cuba made the transition from a highly unsustainable industrial society to a sustainable mixed economy. A new low carbon economics could bring new opportunities. One example of this comes from Kinsale, a town of some 8000 residents, near Cork in Ireland. The 'Kinsale energy descent action plan' explores the de-carbonisation of education, healthcare and other key aspects of delivering well-being in annual steps to 2021 (Hopkins, 2005).

We now have a chance to change everything, because everything must be changed. Facing our challenges could allow us to create the kind of world we actually want to live in. The choice is clear: On one hand, if a minority of powerful nations continue to favour an economic system under-pinned by centralised nuclear and fossil fuel based technologies with inherently vulnerable supply lines, they will need to protect it with a huge world-wide police force at enormous expense and risk to all our civil liberties. On the other hand, if we all begin a shift to a decentralised world economy based on equitable and efficient use of renewable energy sources, and re-localised supply systems, we can create communities that no terrorist organisation can easily threaten and, perhaps more importantly, which threaten no one else.

But this means using the time and the oil, gas and coal we have left to their very best effect, by using them very much slower, so they will last over many centuries, and are burned at a rate which will allow us to meet our climate change targets. If we wait until the challenge is really upon us before becoming serious about developing the solutions, in the ensuing chaos we may no longer be able to muster the resources required. Although we must act quickly, we must also bear in mind that we will be travelling in unfamiliar territory. Information will come from a broad spectrum of (quite possibly biased) sources, so we must always include: full life cycle costs, energy return on energy invested and all externalities. We must remain methodical, clearly distinguishing between total primary energy and total electricity, power and energy, peak and average loads and use common energy units throughout. An ill-informed decision at this point of time could very well come back to haunt us.

In their report *The Role of Nuclear Power in a Low Carbon Economy* the UK Sustainable Development Commission express concern that a new nuclear power programme could divert public funding away from more sustainable technologies that will be needed regardless, hampering other long-term efforts to move to a low carbon economy with diverse energy sources. In fact the majority view of the Sustainable Development Commission is that there is no justification for bringing forward plans for a new nuclear power programme, at this time, and that any such proposal would be incompatible with the Government's own Sustainable Development Strategy (SDC, 2006).

Our utilisation of energy has changed a great deal over the last 50 or 100 years, and there is every reason to suspect it will continue to change, we must, however, ensure that the changes are made consciously and with an eye for the future. The energy is out there, we have the renewable resources, the steel, the skills and the 'post north sea' offshore technologies. It is visionary politics which is now required, uniting the various factions in a radical new initiative, echoing the enthusiasm and fervour of the Apollo programme of the 1960s, linking international agreements, economic policy, technological innovation, academia, R&D, public education, international trade and the globalised media into a unified response.

We are certainly near a 'tipping point' in terms of climate change and peak oil. We are perhaps upon the tipping point in terms of global equity. It seems possible that we are also near a 'tipping point' in terms of an emerging collective consciousness and readiness to act in unison. I justify my optimism with the thought that when we are challenged and realise our necks are on the line, we humans can be inventive, co-operative and highly adaptive. All that is required is that the truth be told.

In a nutshell, it is 'exam time' for humanity, if we can work together and agree on the right choices we can all pass, if we can't, we may very well all fail.

References and further reading

ASPO (2006) The Association for the Study of Peak Oil & Gas, Box 25182, SE-750 25 Uppsala, Sweden, www.peakoil.net.

Heinberg, R. (2004) *Powerdown*. Clearview Publications, USA, DVD format: published by www.globalpublicmedia.org.

Hopkins, R. (2005) *Kinsale Energy Descent Action Plan*. Kinsale Further Education College, Kinsale, Ireland. See also www.transitionculture.org.

Meyer, A. (2000) *Contraction and Convergence*. Schumacher Briefings 5, Green Books, Dartington. See also Global Commons Institute http://www.gci.org.uk/.

NEF (2006) *The UK Interdependence Report: How the world sustains the nation's lifestyles and the price it pays*. New Economics Foundation/Open University. See also 'Well-being and the environment – achieving "One Planet Living" and maintaining quality of life'. New Economics Foundation/WWF.

SDC (2006) *The Role of Nuclear Power in a Low Carbon Economy*. Sustainable Development Commission position paper, London.

Part III The Future UK Energy Mix: Strategic Issues

6

Time for a Fresh Look at Nuclear?

Stephen W. Kidd

Introduction

Nuclear issues have suddenly come back to the fore after a period where they were largely out of the public eye. There was previously a general acceptance, in the Western world at least, that nuclear power would gradually wither away, as existing stations closed and few new ones started up. Nuclear was mainly in the news when weapons proliferation concerns surfaced, but nuclear power itself was very much out of the spotlight. It was widely regarded as too dangerous (in the aftermath of Three Mile Island and Chernobyl), too expensive (many stations suffered delays in their construction period which destroyed their economics) and facing unavoidable difficulties in disposing of hazardous wastes, not to mention the risks of illicit use of nuclear materials by terrorists and others.

This feeling that nuclear power has past its prime, indeed suffering a slow and elongated death, was also common in what is called by its critics the 'nuclear industry'. Opponents like to give the impression that there is a powerful industry battling against them, whereas in fact it has always consisted of a set of not-so-powerful companies involved in various activities throughout the nuclear fuel cycle, but without the 'clout' of the oil or other major industrial sectors. Loosely linked by limited common bonds, pessimism about its prospects was certainly very evident within its own ranks, as the weight of activity moved away from the commissioning new reactors towards waste management and the decommissioning of old plants.

But now there is a revival of interest, caused by a number of factors including rising fossil fuel prices, doubts about energy security from dependence upon supplies from politically unstable locations and worries about greenhouse gas emissions from fossil fuel burning. This has brought the inevitable backlash from those opposed to any further investment in nuclear. As many of the arguments are the same as those used in the past, it is worthwhile looking firstly at how such strong attitudes built up against nuclear and how those on the other side attempted to rebut them. We can then bring things up-to-date to see if anything essential has changed. Is there really still so large a gap between the two sides to prevent any compromise? If not, what can be the terms?

Early days

Opposition to anything to do with nuclear is central to the Green creed – indeed, it is an issue such groups have used as a unifying force among their often disparate members. The roots of anti-nuclear sentiment clearly stem from the links between civil nuclear power and nuclear weapons. The devastating effects of the atom bombs dropped on Japan in 1945 and the fears induced by the subsequent nuclear arms race between the superpowers is something that the civil nuclear sector has never completely succeeded in casting off. The campaigns against nuclear weapons in the 1950s and 1960s involved an entire generation of young people from many political persuasions, who found a common cause around which to rally. The idealists who hoped that nuclear weapons could be removed from the face of the Earth have been sadly disappointed – perhaps not surprisingly, as it is difficult to 'un-invent' a proven technology. But it is arguable that the movement has actually been rather successful on the basis that there has been no subsequent use of nuclear weapons after Japan in 1945, the number of countries possessing them has hardly increased and the quantity of warheads held by the major powers has been reduced by arms limitation treaties. Testing of weapons is now also greatly constrained by treaty.

The civil nuclear power industry cannot escape its obvious origins within the military programmes and there are links still evident today – from the civil side, the most beneficial of these is the down-blending of Russian highly enriched uranium (HEU) from former warheads, which is supplying reactor fuel to satisfy around 10 per cent of current US electricity requirements. Many of those formerly marching against the bomb still have deeply-held convictions against any use of nuclear technology; indeed in extreme cases, even the beneficial applications in medicine and agriculture. Nuclear power stations provide a very obvious symbol of something people are, at best, very suspicious of and, in other cases, strongly opposed to. Major incidents, such as Three Mile Island and Chernobyl, are seized upon by opponents as evidence of society's foolishness in playing with the nuclear devil, but perhaps as important is the collation and clever presentation of masses of evidence of seemingly minor incidents and weaknesses in the nuclear case.

Weaknesses and denial

Supporters of the civil nuclear industry argue that this is all rather unfair. They point to the barriers between the military and civil sides of nuclear and claim that the connections made by opponents are illogical. For example, not many people object to the widespread expansion of civil aviation on grounds that planes can also be designed as formidable fighting machines able to deliver death and destruction. By any scientific evaluation, the industry's safety record is excellent and studies show that the external costs of nuclear are minor when compared with other electricity-generating technologies. Thousands of deaths in coal mining each year, explosions at gas terminals and devastating floods when hydro dams are breached receive a fraction of the publicity accorded to even minor nuclear incidents. Those in the industry know that journalists like an easy story and nuclear provides this only too readily, as it is impossible to completely avoid every minor incident. Each of the main arguments used against nuclear, such as safety, waste management, risks of proliferation and economics have been rebutted as far as possible, yet the general anti-nuclear sentiment has been very hard to shift. People who live near nuclear power stations are usually highly supportive, on the basis that they provide stable well-paid employment and few problems, but there remains a widespread view elsewhere that nuclear is a risky option and its proponents merely acting out of self-interest.

Industry difficulties

If the industry's case is so strong, why has it not been more successful at rebutting its opponents? There are four main reasons – poor communications, the sheer number (if not the quality) of arguments utilised against it, the deep emotional currents that often swamp consideration of the facts in people's minds, and finally the changes in the political process in key countries.

The civil industry's early communications with its stakeholders (to use modern parlance) was undoubtedly appalling – indeed this remained the case until comparatively recently. Arrogant scientists and engineers would address audiences and the media as if they were children – basically saying, 'We've developed this marvellous new technology for you, so you'd better go out and use it. Just do as I say!' Memories of 'too cheap to meter' are frequently brought up and maybe exaggerate what was generally said, but the obvious potential costs of nuclear were certainly dismissed and only the benefits given any credence. The other arrogance was to suggest that nuclear could eventually dominate the energy world, on the basis that fossil fuel supplies would soon run out and become uncompetitive. It has taken a long time for the industry to live all this down, with attempts to use modern communication techniques taking time to bear fruit. This is now gradually happening, but it is proving a long haul to overcome the legacy of the past. It is not a matter of slick industry salesmen in sharp suits now replacing the well-meaning but incompetent scientist-communicators of the past, but more of having eager willingness to engage with all groups of society and patiently explain both pros and cons of nuclear and other technologies.

The industry argues that each of the key arguments used against nuclear technology has very little merit. In each case it may well be realistic to persuade 95 per cent of the people on this. Or alternatively to persuade everybody with a 95 per cent degree of certainty in his or her mind. Yet the residual 5 per cent remain very important, because they are additive. The 5 per cent doubters in the population on one aspect (for example, risks of nuclear proliferation) may be an entirely different group from those concerned about another issue (maybe plant economics). So with several separate arguments used against nuclear, it is not difficult for opponents to achieve a large number of doubters in the population – maybe not 50 per cent but at least a significant minority. Alternatively, the 5 per cent elements of doubt in any individual's mind on each issue are similarly cumulative. Lots of 5 per cents begin to add up to the extent that many people will say, 'Well, there has got to be something wrong with this technology, as so many little things can go wrong – so lets use something rather simpler'. This 'wearing-down' process accounts for much of the anti-nuclear movement's success – no matter how many arguments are rebutted, there always seems to be another one.

The industry has put a lot of effort into presenting the facts about nuclear power and other power-generation technologies, through establishing good websites, providing media interviews and addressing conferences and other interested audiences. This has certainly helped counter some of the more unreasonable claims of the anti-nuclear movement, but it has not been enough. More third party advocates have been needed but these have only comparatively recently emerged, notably some formerly identified as leading environmentalists. The bigger problem, however, is that the debate cannot be answered solely by reference to the facts. It is conceivable that both sides can agree the key facts, but the interpretation of these and their meaning can differ appreciably. This is because of different views on risk-taking and the values one ascribes to aspects of the world – in other words, we are in the area of welfare or normative economics. Careful examination of and attempting to agree on the facts will undoubtedly help, but it cannot resolve the matter. For some people, a 1 per cent chance of a nuclear accident in the United Kingdom over the next 100 years causing 100 deaths may be completely unacceptable, but can be taken in their stride by others who know of the extent of coal-mining deaths each year in China.

It is clear that nuclear power needs top-level political support to prosper in any country. Not, note, financial support, but at least the establishment of a reasonable licensing and regulatory regime, defending the interests of all parties, plus clear policies on aspects such as used fuel management and plant decommissioning. Uncertainties on these are fatal to a technology requiring heavy up-front investment followed by many years of operation to recoup these costs and then make a profit. Yet such support has become hard to win as politicians have generally become more reactive, responding to focus groups and the like, rather than strong conviction-led leaders. They know that nuclear is an issue that gets a small percentage of the population very excited, either pro or con. So if the government comes out strongly in favour of new reactors, for example, it is likely to lose the votes of all those fiercely opposed to nuclear, irrespective of other considerations in the next election. These votes could be crucial in a tight ballot; so nuclear is, for them, a dangerous issue. Thus it tends to be swept under the carpet through fence-sitting, putting off energy revues until later and so on. We have only recently begun to see the reversal of this, particularly with the Bush Administration's strong support for nuclear in the US, but it may take greater general public acceptance before other politicians put their necks on the line.

Differences today?

It is interesting to assess whether the historic arguments used against nuclear are today any different and what has happened in the period since to influence their validity.

In essence, there seems little change in the views of the antis. Allegations of poor plant safety, the dangers of radiation, risks of weapons proliferation, waste management difficulties, poor economics, high costs of plant decommissioning, transport risks and the inadequacy of long-run uranium resources remain the main attack points. Yet on most of these aspects, nuclear proponents can fairly claim that the industry has performed very well and the risks have been overstated. The safety record is excellent, both at the nuclear plants, other nuclear facilities and in transport, while economics have been transformed by cutting costs, better plant performance and substantial fossil fuel price increases. The two areas where the industry faces the greatest public perception challenges today are on used fuel management and potential weapons proliferation. The delays to the establishment of waste repositories have been damaging; indeed, the concept of putting the fuel in deep repositories for thousands of years remains under some challenge. Although nearly all the stories of nuclear trafficking prove irrelevant, the recent proliferation of enrichment technology in Iran and North Korea has increased worries. The Treaty on the Non-Proliferation of Nuclear Weapons (NPT) and the attendant International Atomic Energy Agency (IAEA) safeguards provisions have actually worked very well in practice, but they remain imperfect and are subject to much further discussion at present.

What has also come more strongly into focus is a detailed discussion of what is meant by a Green technology, within the wider concept of sustainable development. Nuclear proponents have not been slow to promote its clean-energy attributes on account of the almost absence of carbon emissions in the nuclear fuel cycle and in subsequent power generation. Attempts by opponents to argue that nuclear is not so clean on carbon or doesn't actually provide any net energy addition, can easily be shown to be nonsense – the worst sort of Green propaganda. Yet there is plenty of room for informed debate on this overall subject. For example, is wind-generation technology really so advantageous, when so much metal and plastic is used in the turbines and many people object to them on aesthetic grounds? Nuclear was excluded from specific advantage in the Kyoto Treaty but is now receiving competitive benefit from emissions trading, which penalises heavy carbon-emitting technologies. Is this enough – should it not receive more if avoiding carbon emissions is such an important societal goal?

Common ground?

Can there not be some common ground where nuclear proponents and the sceptics can agree? Probably not, in the hardened 'cases' on both sides. For one thing, value systems on each side are very different. Nuclear opponents in the 'Green Movement' are generally against many trends apparent today, such as globalisation and centralised decision-making on important matters such as energy. Quite apart from any environmental objections, they see nuclear power programmes as embodying all they hate about the modern world, whereas they would like to see a move towards smaller, decentralised energy supply based on renewable technologies. There has accordingly been a general tendency in the media to promote nuclear and renewables as competitive, non-carbon-emitting technologies, but cannot they be seen as complements? Nuclear is well suited to covering base-load electricity requirements while renewables can add substantial power increments on top. Where are the areas where it may be possible to agree?

The first of these is surely to recognise that each power-generating technology has significant costs and benefits. It is therefore highly unlikely that any should completely dominate energy production and that a mix of solutions is the optimum in nearly all circumstances. Renewables contain their own individual mixes of advantages and disadvantages, but some of their proponents unfortunately display the same arrogant, blinkered thinking displayed by many nuclear advocates in the 1950s and 1960s. For both OECD nations and the developing world, a mixture of generating options is surely appropriate, determined by resource endowments, geography, energy security and similar considerations. To rule out any option through ideology is not appropriate.

Secondly, determining the generating mix should, as far as it is possible, be informed by reference to the facts. We should therefore hear no more of the more ridiculous arguments against nuclear, such that it provides no net incremental energy addition. Can France, with an 80 per cent nuclear share in its electricity generation mix, really have got things so wrong? It is clear that the public still needs better information on energy matters as years of cheap fossil fuels have induced complacency. Yet there is no need for advocates of any energy solution to engage in 'knocking copy' against the others. The expected growth in world electricity demand, perhaps doubling by 2030 according to the International Energy Agency, leaves plenty of room for substantial growth for everybody.

Thirdly, it should be common ground that emerging energy technologies should receive public subsidies in order to allow them to develop. Nuclear received substantial public backing in the past and renewables deserve the same today. It is becoming increasingly clear that the first new nuclear units to be built will not need financial subsidies, as the economics now look sound, assuming that investors can take a long-term view (WNA, 2005). The key requirement is for the public authorities to develop a clear regulatory environment and develop national policies on waste management and decommissioning – new plants can then set aside appropriate funds to cover these future liabilities. Nuclear power does not necessarily have, in the future, to be a creature of 'big government' and need not 'crowd out' a desirable rapid expansion of renewables. Providing cheap and largely carbon-free base load power 24 hours per day is a sound basis for any electricity system, with other generation options supplying the balance. To some extent, there may be a competition for limited government attention and funding, but if low carbon energy solutions are important, it should not be impossible for government to encourage nuclear and renewables to expand simultaneously.

Conclusions

It makes good sense for the advocates of both nuclear and renewables to try to throw off the baggage of the past and move forward together. Neither is going to go away and so friendly coexistence would seem to be a good policy. If a reduction in carbon emissions is needed, nuclear technology is available today. That it will take some time to build new nuclear plants argues for streamlining the regulatory regime as far as is compatible with meeting reasonable requirements for public review. The share of renewables in world electricity is also clearly capable of rising substantially, but it will similarly take time for this to occur. The long-term energy future is very uncertain – if we move to systems based largely on hydrogen rather than hydrocarbons, there are good possibilities for both nuclear and renewable technologies. Yet decisions should be made after careful presentation and discussion of all the facts, and no side has any true interest in these being hidden or obscured.

References and further reading

WNA (2005) *The New Economics of Nuclear Power*, World Nuclear Association, London: WNA *website http://www.world-nuclear.org*.

7

Nuclear Power and Renewables in the UK: Can We Have Both?

David Elliott

Introduction

It is sometimes argued that we can and should have both nuclear power and renewable energy. Implicit in this view is the belief that they can be compatible and are not in competition. This chapter attempts to explore that contention in the UK context, by looking at how the two technological development programmes have interacted in the past and how they might interact in the future.

The analysis in this chapter assumes a major commitment to energy efficiency, as a common and crucial feature for any sustainable energy future (Elliott, 2004). So demand/efficiency issues are not explored. Instead the focus is on the non-fossil supply options – nuclear and renewables – and potential conflicts between them.

The story so far

Historically, in the United Kingdom and elsewhere, the nuclear industry has enjoyed major funding allocations in terms of support for research and subsidies for operation, while the renewables have tended to be sidelined. For example, between 1974 and 2002 nuclear fission received around 47.3% of the overall R&D funding in the IEA countries, and nuclear fusion around 10.5%, while renewables only received about 8.1% (IEA, 2004). A study of Federal Energy Subsidies by the US Renewable Energy Policy Project in 2000 found that the US government had by that time spent approximately $150 billion on energy subsidies for wind, solar and nuclear power, but that 96.3% of this had gone to nuclear power (REEP, 2000). Globally it has been calculated that subsidies for energy overall were around $235 billion in 2004, but renewables only received around 7% of this (IIED/NEF, 2004). The WISE/NIRS Nuclear monitor issue 630/31 notes that the US nuclear energy sector received $15.3/kWh in the first 15 years of the development of nuclear power (1947–61) whereas wind energy received just $0.46/kWh in its first 15 years – 30 times less.

In the United Kingdom, although the level of funding support for nuclear has in general declined in recent years, it still attracts significant state funds, including state cover for clean-up costs and other nuclear liabilities. For example, the UK government initially earmarked around £48 billion for site decommissioning and clean up work over the decades ahead, although the estimated cost was subsequently revised to £56 billion, and then to around £70 billion. However, leaving these very large 'historical liabilities' costs aside, in terms of support for new generation technology, renewables are now beginning to gain on nuclear (see Table 7.1).

The fact that renewable funding has now increased while nuclear funding has declined is sometimes seen as evidence that there is, in effect, a 'fixed sum game': with inevitably finite amounts of cash being available, they are in competition, as one prospers the other declines and their relative fates are intertwined. The reality is more complex. For example, in the United Kingdom, overall Department of Trade and Industry (DTI) energy expenditure has begun to expand, albeit from a low level, so that nuclear and renewables have both been able to obtain more funding, although the renewables have gained more, while nuclear has fallen back dramatically from the high levels of R&D funding it was given in the past. For example, in the 1970s, nuclear was receiving typically around £200 million per annum for R&D.

Even so, most people in the renewables community still see nuclear as a rival, a view strengthened by the fact that the nuclear industry has sometimes been disparaging of the contribution that renewables could make. It is also significant that many of the early renewable energy pioneers saw renewables as an alternative to nuclear power and this implicit hostility to

Table 7.1   Department of Trade and Industry expenditure on energy (£ million)

	2000–2001 Outturn	2001–2002 Working	2002–2003 Plans	2003–2004 Plans
Nuclear (excluding UKAEA liabilities)	21.8	49.7	52.7	57.8
Renewable and novel sources of energy	11.0	17.7	18.5	19.0
Renewable capital grants scheme	0.0	13.0	45.0	47.0

Source: http://www.dti.gov.uk/expenditureplan/report2004/

nuclear has continued to form the wider debate over energy and environmental futures. At the very least, few people in the renewables community seem to have wanted to risk association with nuclear power, given the latter's very poor public image.

It is interesting in this context that, at various points in the past, the nuclear lobby has sought to make tactical or strategic alliances (variously against coal, or against gas) with the renewable energy community. Each time its advances have been spurned, sometimes quite vehemently, often reflecting a lack of trust in the nuclear lobby's motives. For example, in an editorial in *Wind Power Monthly*, 5 June 2005, Lyn Harrison, the magazine's editor, commented: 'In a true wolf-in-sheep's clothing trick, the nuclear lobby pours forth woolly words on "partnerships" with renewable energy, while savaging wind behind the scenes'. Perhaps this is evidence of inflexibility on the part of the renewable energy community, after all a partnership with nuclear might have some strategic advantages. Or perhaps it is just a conviction that nuclear is the technology of the past and renewables the technology of the future.

For many people in the renewable energy community it is simply obvious that, quite apart from the economic, safety, security, waste and proliferation issues (some of which are discussed later in this book), large capital-intensive centralised power plants have no place in the future of dispersed locally embedded generation, whether in the developed or developing world. In particular, nuclear power is seen as unsuitable for underpinning appropriate patterns of economic development. For example, it is unlikely to be relevant to the 2 billion or so people who currently do not have access to electricity and are unlikely ever to be within reach of power grids. By contrast many renewables are well suited to meeting their needs, being smaller scale, often modular and using local energy resources, thus avoiding the cost of providing grid links (WADE, 2004).

Clearly there are disagreements about what should be the appropriate pattern of economic development, but many observers now agree that industrialisation on the Western model is not environmentally sustainable. Instead they argue that there is a need for new patterns and therefore new technologies. Some critics might be willing to negotiate on the inclusion of smaller safer nuclear plants in this package, if, in terms of responding to climate change, there was absolutely no alternative. But to most that sounds like a very long shot. There are, they claim, easier and better technological options, and it is frustrating for renewables enthusiasts when these are constrained by what often seems a perverse commitment to a flawed technology.

The provision of larger grants to renewables may help reduce this hostility and resentment, but equally the experience of the long and sometimes bitter struggle by renewables enthusiast to try to get recognition for their vision in the face of resistance from powerful pro-nuclear lobbyists is unlikely to be easily forgotten. The view is still widely shared that if nuclear power is given an opportunity it will once again absorb all the funds available at the expense of renewables. Some environmental activists feel that evidence for this view is provided by recent events in Finland, a country with enviable renewable resources, which has, however, decided to build a new nuclear plant. Activists claim that, despite initial governmental promises otherwise, following the new commitment to nuclear, support for renewables has been reduced (Nuclear Monitor, 2004).

This type of outcome need not be inevitable. After all, technically, it seems possible for nuclear and renewables to co-exist and even to reinforce each other to some extent. Although they cannot easily load follow (nuclear plants can't easily be run up and down to match changing power demand) nuclear plants can provide firm base-load power, which might compliment renewables. However, there are different engineering and system development philosophies involved. Nuclear plants are large, centralised units, feeding power to remote consumers via the national super grid. Some renewables by contrast are small scale, providing power directly to local users, including micro power units (e.g. photovoltaic (PV) solar, wind) used at the domestic scale. If the idea of locally embedded generation by small distributed plants and micro-generation by users continues to gain support for both electricity and heat supply, then large grid-linked plants may begin to look out of place. For example, far from ensuring diversity in terms of the range of supply option, emphasis on large centralised nuclear plants might be seen as seriously constraining the development of a more robust energy system based on a diverse range of smaller geographically dispersed energy inputs of heat and power (Greenpeace, 2005).

There would thus seem to be real conflicts in approach. In its report on nuclear power the Sustainable Development Commission argued that

> reliance on centralised supply may exacerbate the current institutional bias towards large-scale generation, and the reluctance to really embrace the reforms necessary to ensure a more decentralised and sustainable energy economy.

The lack of flexibility, or lock-in, associated with investment in large-scale centralised supply like nuclear power is also a concern. This relates to the issue of sunk costs. A new nuclear programme would commit the UK to that technology, and a centralised supply infrastructure, for at least 50 years. During this time there are likely to be significant advances in decentralised technologies, and there is a risk that continued dependence on more centralised supplies may lock out some alternatives. Decentralised supply is generally more flexible because it is modular, and can adapt quicker and at less cost to changed circumstances. More locally-based energy provision may also be conducive to the sustainable communities agenda, a key part of the UK Government's Sustainable Development Strategy. (SDC, 2006)

Equally, a significant nuclear element may not be operationally compatible with a large renewables programme. For example, if renewables became a major element in energy systems, then there would be a need for some flexible back up plants to compensate for the intermittency of some renewables sources. This is not a major problem at low and medium levels of input, but it will add to the cost as the proportion of renewables increases, although, to put it in the context of our comparison, the extra costs are not as large as the cost of providing waste storage facilities for spent fuel from nuclear plants. Longer term backup may be provided by biomass-fuelled plants, but initially it will have to be fossil fuel-fired plants – small flexible gas turbines which can load follow. Nuclear plants are large and inflexible and cannot easily be used to load follow, so they may find themselves marginalised, if and when the energy economy becomes increasingly based on small distributed renewables (Everett, 2005). Equally, if there is a large nuclear component, then it would be hard for wind power to expand since, to ensure economic operation, nuclear plants have to be run continuously, so that when demand is low, any excess wind plants would have to be shut down. *Wind Power Monthly* has argued that 'with 20% nuclear on a power system, only 10% wind can be accommodated before economic penalties cut in. At 40% nuclear, wind is out of the picture' (Wind Power Monthly, 2006).

What next?

Reasonable progress has been made on developing and deploying renewables in the United Kingdom. However, it is increasingly argued that the pace of expansion of renewables will be insufficient to meet the greenhouse gas reduction targets the United Kingdom has set, and that one solution is to switch funding back to nuclear. Many in the renewables community view this possibility with distaste, seeing it as very short-sighted. The counter view is that it would be better to give renewables a chance to show what they could with proper funding on the basis of the conviction that the funding supplied so far has not been sufficient. Moreover, many people in the wider environmental movement, including members of Friends of the Earth and Greenpeace supporters, are also strongly opposed to the idea of a return to nuclear, and, if that is pursued, we might see a return to the direct-action anti-nuclear demonstrations of the 1970s. Indeed they have continued in Germany, in relation to nuclear waste shipments, and there is nowadays a wide ranging and militant anti-globalisation movement which might take up the anti-nuclear cause, especially given the global weapons proliferation implications.

Quite apart from the eco-activists, most public opinion polls in recent years have shown that while the majority of people in the United Kingdom do not want nuclear power, they support renewable energy. For example, a market & opinion research international (MORI) poll in 2002 found that 72% of those asked favoured renewable energy rather than nuclear, while a National Opinion Poll carried out in the same year for the Energy Saving Trust found that 76% believed that the government should invest time and money in developing new ways to reduce energy consumption, 85% wanted government investment in 'eco-friendly' renewable energy (solar, wind and water power) and only 10% said the government should invest time and money in building new nuclear plants.

However views may be changing, in response, for example, to the a claim made increasingly by the nuclear lobby that nuclear power can make a major contribution to reducing greenhouse gas emissions, and media reports suggesting that we may face power shortages without it. A MORI poll carried out for the Nuclear Industry Association in 2004 found that, whereas in 2001 only 19% in their sample had been in favour of a nuclear replacement programme, in 2004, 30% said they would support the building of new nuclear power stations to replace those stations that are being phased out, while, although objectors were still in the majority, those saying that they would definitely oppose a replacement programme, had fallen from 67% in 2001 to 34% in 2004 (MORI, 2004).

A new debate over which way to go is clearly unfolding, although care has to be taken with public opinion surveys, the results depends crucially on the questions asked, and some surveys have continued to indicate that support for nuclear power was very conditional. An Ipsos MORI survey of 1500 people carried out in 2005 jointly by the Centre for Environmental Risk and the Tyndall Centre for Climate Change at the University of East Anglia, found that while 54% might be prepared to accept new nuclear plants if they were shown to help tackle climate change, 78% thought renewables and energy efficiency were better ways of tackling global warming.

Future prospects

The nuclear debate has been revived in recent years in part because of the claim that nuclear power could play a role in responding to climate change. This view has been promoted for some time by the nuclear industry (Donaldson et al., 1990), but it only reached a wider audience following the increase in concern about climate change and the intervention, in 2004, of a handful of prominent people, most notably the environmentalist James Lovelock (Lovelock, 2004). Their claim was basically that the problems of nuclear power were small by comparison to the threat of climate change, and that renewable energy could not be relied on as an alternative.

This claim can be challenged in several ways. Firstly, there is the basic issue of whether nuclear power could actually make a significant long-term contribution, given that reserves of high-grade uranium are limited. Indeed, depending on the rate of uranium use, it has been argued that there may not even be a significant medium-term future (Mobbs, 2005), although clearly this view is not shared by the nuclear lobby. For example, it is argued that new reserves will be found and that, longer term, uranium could be extracted from sea water, despite the very low concentrations (Price and Blaise, 2002).

However, that leads on to a more general problem. If nuclear power was expanded on a major scale globally, then recourse would have to be made to lower and lower grade ores and ultimately unconventional sources like seawater. Since most of the energy required for fabricating and enriching fuel would still come from fossil-fuelled plants, there would be increasing levels of carbon dioxide production, thus undermining the ostensible advantage of the nuclear option – no direct greenhouse gas production. Moreover, it has been claimed that, at some point, if a very large nuclear programme was launched globally in response to climate change, the emissions associated with the fossil fuel-derived energy used for reactor fuel production would be larger than those produced if the fossil fuel was simply used to generate electricity (van Leeuwen and Smith, 2005).

This view has been challenged by the nuclear lobby (WNA, 2005), and the dispute over the data and the assumptions used in this type of energy/carbon analysis led the House of Commons Environmental Audit Committee to suggest, in its 2006 review of UK energy policy, that the Royal Commission on Environmental Pollution be asked to review the issue (EAC, 2006). Certainly there is room for debate. For example, it has been argued that, when and if the global nuclear programme expanded, more of the energy needed for fuel production could be provided by nuclear plants and other low carbon sources (SDC, 2006). While this is true, there would arguably be diminishing returns in terms of energy costs, and diminishing uranium reserves. It would, arguably, make more sense in energy resource terms to invest energy in building renewable energy technology, since this is not resource limited – the fuel is free and will be available indefinitely.

There might of course be technological breakthroughs. The development of fast breeder technology would extend the uranium resource, albeit with a range of economic, security and environmental costs. However, it would not solve the resource problem in the long-term (Mobbs, 2005). Nuclear fusion might at some stage provide a new energy option and would have much longer resource lifetime, although again there could be a range of impacts and the eventual generation costs are unknown (Pearce, 2006). For the moment, the only nuclear options available are the various conventional fission reactor designs, including the new, and as yet unproven, upgrades of the pressurised water reactor, e.g., the Westinghouse AP 1000 and the European Pressurised Water Reactor, and longer term, possibly versions of the high temperature 'pebble-bed' reactor prototypes being developed in China and South Africa.

The second main challenge to the nuclear claim concerns the assertion that it is more desirable and indeed more viable than renewables for the immediate future. For example it is sometimes argued that nuclear power is, or will be, cheaper than the renewables. Certainly the nuclear industry has claimed that new nuclear technologies will be more competitive. It is hard to project prices into the future, but a study by the Performance and Innovation Unit carried out for the UK Cabinet Office in 2002 found that the long-term prospects for most renewables looked quite good compared to nuclear power (see Table 7.2).

Another area of debate is over the relative social and environmental impacts of nuclear and renewables. This touches on a wide range of often contentious issues (Elliott, 2003a). The most obvious concern the risks of major accidents, which have particularly worrying long-term implications for nuclear, as is discussed in Chapter 8, but are also an issue with large hydro. Similarly, there is the risk of terrorist's attack, which mainly seem confined to nuclear (as is discussed in Chapter 10), but could conceivably also be relevant to large-scale hydro. More generally, there are the heath and safety implications associated with the normal operation of the various energy options. For example, in relation to nuclear, as is discussed in

Table 7.2 Cost of electricity in the United Kingdom in 2020 (in pence/kWh)

On-shore wind	1.5–2.5
Offshore wind	2–3
Energy crops	2.5–4
Wave and tidal power	3–6
PV Solar	10–16
Gas (CCGT)*	2–2.3
Large CHP**/cogeneration	under 2
Micro CHP	2.3–3.5
Coal (IGCC)***	3–3.5
Nuclear	3–4

*Combined Cycle Gas Turbine, **Combined Heat and Power, ***Integrated Gasification Combined Cycle.
Source: Performance and Innovation Unit, 'The Energy Review', UK Cabinet Office, 2002.

Chapter 8, there are concerns about the impacts of low-level radiation releases from nuclear facilities and the impacts of long-term waste repositories. Equally, there are concerns about the significance of the visual impact of renewables like wind farms and the land-use implications of energy crop plantations (Elliott, 2003b). Making meaningful comparisons between these various impacts is hard. In a recent full life cycle analysis of energy systems the World Energy Council (WEC) warned that, while in terms of greenhouse gas emissions 'renewable fuels and sources and nuclear compare favourably, some of the externalities cannot be covered by the Life Cycle Analysis methodology' – for example, long-term impacts from nuclear waste releases. These it says 'must be addressed within the political process' (WEC, 2004).

Finally, there is the issue of whether nuclear and renewables can be expanded significantly to meet the challenge of climate change, and also the question of whether other technologies may also play a role – it may not be a simple matter of 'nuclear versus renewables'.

At present nuclear power provides around 19% of UK electricity, but this will decline as plants are retired, so that, unless policies change, according to the DTI's initial energy review consultation paper, issued in January 2006, nuclear power's contribution would drop to 7% by 2020, with only Sizewell 'B' operating from 2023 onwards, supplying less than 3% of UK electricity. The nuclear industry argues that we should 'replace nuclear with nuclear', whereas, at present, the plan, as outlined in the 2003 White Paper on Energy, is to try to ramp up renewables to around 20% of electricity by 2020 (DTI, 2003). This seems a reasonable target – Denmark has already achieved it, just from wind. Given its enviable renewable resource, the United Kingdom could actually to do better – if the funding was available. By 2006, Scotland had already reached 16%. In parallel, the wide adoption of CHP (the United Kingdom has a 10 GW by 2010 target) should be making more efficient use of gas, so that, in effect, assuming a major commitment to energy efficiency on the demand side, net carbon emissions can continue to fall in line with the Kyoto targets.

There is of course the risk that the White Papers' programme of reliance on renewables, plus CHP and efficiency, will not deliver its carbon mitigation targets in time, or provide sufficient energy security in the meantime. Certainly each of these energy options could do with much more support. However, if it is felt that we need an additional interim element to provide security while still avoiding emissions, then it has been argued that, rather than returning to nuclear, attention should be given to the idea of enabling the continued use of fossil fuels via carbon dioxide capture and underground sequestration. There is space in undersea oil and gas wells to store several decades of emissions from UK power stations, and although the costs are likely to be high, this could provide at least an interim option.

However, as with nuclear, there is the risk that funding will be siphoned off into this option at the expense of renewables (and nuclear), and the fossil fuel industry would no doubt like to see this option expanded into the future by using as yet unopened undersea saline aquifers, which may provide a much larger storage volume. However that is more speculative, and for the present, in the UK context, the sequestration option seems only to be one that will be viable for a few decades worth of output from fossil-fuelled plants. Moreover, it seems unlikely that many existing UK plants would be worth retrofitting for CO_2 collection. More likely would be investment in a few new gas-fired plants (BP has proposed one in Scotland feeding CO_2 to the Miller field) or coal-fired integrated gasification combined cycle turbines (IGCCT), built close to shore with links to suitable undersea reservoirs. Nevertheless, that could provide one initial route to generation of hydrogen from fossil fuels (with the CO_2 produced being stored), and, although expensive, this could help establish the beginning of a hydrogen economy which could subsequently be supplied by hydrogen produced using electricity from renewable energy sources.

Overall, carbon sequestration, coupled possibly with the use of mine methane (which might support around 400 MW of new generation capacity), could thus provide a useful non-disruptive compliment to the UK renewables programme. Storing some carbon dioxide underground, and making use of methane created underground, is probably less contentious than trying to find secure underground sites for storing increasing amounts of radioactive materials for an indefinite period.

In addition, there is the potential contribution that might be made by individual consumers using domestic scale micro-wind and PV solar, installed on homes to provide power direct, but linked to the grid for top-ups and excess power export. Micro-CHP, using gas-fired Stirling engines, has already begun to lift off in the United Kingdom, and Powergen has estimated that by 2020 at least 30% of UK households could be using a micro-CHP system – there is an annual replacement turnover of around 1.3 million boilers annually. It could be that the use of micro-wind and PV will expand similarly. If, for example, 10 million consumers bought 2 kW systems, that would be 20 GW of extra renewable capacity – which, despite the lower load factors, could produce nearly as much total annual delivered energy to consumers as the UK nuclear plants, given that there would be no long-distance energy transmission losses.

Longer term, further carbon reductions could be achieved by the continued development of renewables, large and small. The Renewable Energy 2004 Conference in Bonn discussed scenarios in which renewables supplied 50% of total global *energy* (heat and power) by 2050 and the United Kingdom is well-placed to achieve something like this and maybe more (Renewable Energy, 2004). For example, the DTI/Carbon Trust's 2004 Renewable Innovation Review suggested that by 2050 the United Kingdom should be able to obtain between 53% and 67% of its electricity from renewables. That would require rapid rates of deployment, but that seems feasible given that, for example, Germany has already installed 18 GW of wind capacity, with a construction rate of between 1 and –2 GW per annum, rising to 3 GW in 2002/2004. This rate of build is not surprising given that, once planning agreements have been obtained and the site prepared, wind turbines can be installed in a matter of days and operational in weeks.

By contrast, given the long planning process and long construction process, it is hard to see how a nuclear power expansion programme could be started quickly, and also hard to envisage one that would deliver large amounts of power rapidly. It is true that the United States managed to achieve installation rates of around 2 GW per annum during the nuclear boom period in the 1970s and France did even better than that, but the UK experience so far has been one of major delays and cost overruns.

A replacement programme in the United Kingdom might involve the construction of, say, 10 new nuclear plants, over perhaps a 10-year period. The industry seems convinced that this would be possible: there is even talk of shorter construction times. However achieving that would require significant financial resources.

That brings us back to our starting point – depending how the nuclear programme was funded, it could undermine or constrain the full and rapid development of renewables. For example, in November 2005, the Energy Minister Malcolm Wicks was quoted as suggesting that nuclear power might be included in the Renewables Obligation (RO) system (FT, 2005). Given that there is around 11 GW of existing nuclear capacity, unless the Obligation was expanded considerably, the result could be that renewables would be squeezed out, although the effect might be lessened if the inclusion in the RO was limited just to *new* nuclear plants. In the event the idea of including nuclear in the RO was not pursued. Instead, the energy review published in July 2006 proposed the idea of enhancing the EU Emission Trading System to provide an incentive framework. Given that, so far, the carbon market has proved to be very volatile, doubts have been expressed about the viability of the EU-ETS as a way of providing the stable levels of extra support that potential investors would seek. But given growing concerns about climate change, the carbon market might settle longer term and provide a stable context for nuclear investment. That could also support renewables. However, even leaving aside the issue of other hidden subsidies and concessions (e.g., on waste management, decommissioning and insurance cover), the enhanced carbon prices would mean that nuclear power would gain advantages it did not enjoy before – not only has electricity from nuclear plants been excluded from the RO, it has also been denied exemption from the Climate Change Levy. The new proposal thus represents a significant shift, and there would be a risk that this new support for nuclear would deflect investment from new renewables.

No doubt the increased funding and support that has been given to renewables in recent years could be seen as having had a similar impact on nuclear powers' prospects. As noted earlier, it may not be a 'fixed sum game', but to some extent the limited expansion of renewables has been at the expense of nuclear. The point seems to be that they are in conflict over resources: they cannot both expand rapidly.

Conclusions

While economic assessments can provide a guide, the choice of new energy options is a matter of technological faith and strategic judgement. The renewables seem a credible option for the longer term, if given appropriate support, but many governments still seem wary of backing them as the main option for the immediate future. Some remain tempted by the nuclear option. Despite the technical and financial problems nuclear power has had so far, they have faith that the nuclear industry can come up with new cheaper and safer technologies. Even so, they are reticent about 'putting all their eggs in one basket', so that the compromise of 'trying for both' has an appeal. But it also has costs. Nuclear technology, as so far developed, is large scale and very capital intensive. This makes it inflexible in terms of rapid deployment. It is also likely to 'crowd out' other options. By contrast renewables are smaller scale, diverse and flexible. However, because of this, it is easy for them to become marginalised, as they have been until recently. In this situation, it is hard to see nuclear and renewables as anything else but rivals. In July 2005, noting that there was talk of restarting the nuclear programme, the House of Commons Environmental Audit Committee commented that there was 'some concern that uncertainty regarding the Government's intentions in this respect might also damage future investment in renewables and energy efficiency', a view that was followed up in their subsequent report. A similar view was taken by the Environment Agency in their submission to the energy review. It said that it was concerned 'about the displacement effect that a large programme of investment in one capital-intensive technology like nuclear may have on energy efficiency, CHP and renewable technologies' (Environment Agency, 2006).

In its March 2006 report, the Sustainable Development Commission was even more forthright. It claimed that 'a new nuclear power programme could divert public funding away from more sustainable technologies that will be needed regardless, hampering other long-term efforts to move to a low carbon economy with diverse energy sources' (SDC, 2006).

Given inevitably limited overall resources, some conflicts seem inevitable, although some may also be due to the history of the two technological areas and the single-mindedness with which the nuclear option has been pursued. In 1999, the Royal Society/Royal Academy of Engineering produced a report on responses to climate change, suggesting a UK non-fossil programme costing $450 million per annum as part of a global programme rising to $25 billion per annum, much of which it seemed to see as going to nuclear power. It argued that 'very large resources are needed' for, among other things, novel nuclear reactors and waste disposal. By contrast, the report suggested that 'more modest resources' could 'accelerate the development of new photovoltaic systems ... provide the systems development needed for offshore wind farms (and) establish the feasibility of wave power' (Royal Society, 1999).

As noted earlier, this imbalance in approach has persisted for many decades. Nuclear has been seen as important and needing major funding, while renewables have been the poor relation. In part, this imbalance has reflected the fact that renewables were new technologies, while nuclear was well established, although it could be argued that this perhaps should imply that it should need *less* funding, not more. However, as we have seen, the pattern has now begun to change, at least in some countries – renewables are beginning to get larger allocations. In principle, this might begin to reduce, some of the historical resentments, although equally, by splitting the funding between nuclear and renewables, an approach which seeks to support both, more or less equally, might risk doing neither well.

On balance, it would seem better to make a clear choice, at least for a specified period, to let one or other have a clear field. Some might say that nuclear power should be given its head, and despite the problems, this approach might be viable – it has been the approach adopted by France until recently. However, a rival view is that, since nuclear has had many decades of extensive funding, it would be reasonable to let renewables have an opportunity to show what they can offer, leaving nuclear as an insurance option for the future, in case it turned out to be needed. An ancillary view is that, since, given the uranium resource limits, in the longer term we will have to rely on renewables, it makes sense to get started on them as soon as possible, rather than detouring back to nuclear for a relatively short period. A rival view is that it will take time to develop renewables, so we should make use of nuclear in the interim. Both policies have costs and risks – on one hand, it will be hard to keep the nuclear option open as an insurance option, and, on the other, there is the risk that, once nuclear power gets re-established as a major option, even if only meant as an interim, the transition to renewables will be stalled. In effect, that is the risk of the compromise option – that a new emphasis on nuclear would delay the development of renewables. If unlimited money was available, then, all other things being equal, perhaps we might consider backing both. But as things stand in the United Kingdom, it would seem that the policy outlined in the 2003 While Paper on Energy makes sense: as far as the non-fossil supply side goes, we should focus on renewables (DTI, 2003).

References

Donaldson, D., H. Tolland and M. Grimston (1990) *Nuclear Power and the Greenhouse Effect*. UK Atomic Energy Authority, Jan.

DTI (2003) *Our Energy Future: Creating a Low Carbon economy*. White Paper on Energy, Department of Trade and Industry.

EAC (2006) *Keeping the Lights on: Nuclear, Renewables and Climate Change*, Sixth Report of Session 2005–06, House of Commons Environmental Audit Committee, March.

Elliott, D. (2003a) The future of nuclear power. In: *Energy Systems and Sustainability* (G. Boyle et al., eds). Oxford University Press, Oxford.

Elliott, D. (2003b) *Energy, Society and Environment*. Routledge, London.

Elliott, D. (2004) Energy efficiency and renewables. *Energy and Environment* **15**(6),1099–1105.

Environment Agency (2006) *Response to the DTI Consultation on the Energy Review.* Environment Agency, London.

Everett, B. (2005) *The UK Electricity Industry: Transforming the Elephant,* NATTA report, Milton Keynes.

FT (2005) '"Green" subsidy considered for nuclear power'. Jean Eaglesham and Christopher, *Financial Times,* London, 26 Oct 2005.

Greenpeace (2005) *Decentralising Power: an energy revolution for the 21st century.* Greenpeace UK, London.

IEA (2004) *Renewable Energy – Market and Policy Trends in IEA Countries,* International Energy Agency report, Paris.

IIED/NEF (2004) *Up in Smoke.* International Institute for Environment and Development/New Economics Foundation, London, Oct.

Lovelock, J. (2004) Nuclear power is the only green solution. *The Independent,* 24 May 2004.

Mobbs, P. (2005) *Uranium Supply and the Nuclear Option. Oxford Energy Forum,* Issue 16, Oxford Institute for Energy Studies, May. See www.fraw.org.uk/mobbsey/papers/oies_article.html.

MORI (2004) *What Do The Polls Tell Us? – Public And MP's Attitudes To Nuclear Energy.* Speech by Robert Knight from MORI Energy Research, at the Nuclear Industry Associations' Annual Conference (which was entitled 'Could the lights go out?'). Dec. See www.mori.com/pubinfo/rk/what-do-the-polls-tell-us.shtml

Nuclear Monitor (2004) NIRS/WISE *Nuclear Monitor.* Issue 615, WISE Finland report, p. 7.

Pearce, F. (2006) 'The return of nuclear fusion', *Prospect* 124 pp. 38–40.

Price, R. and J.R. Blaise (2002) *Nuclear fuel resources: Enough to last?,* NEA updates, NEA News, No. 20.2, Nuclear Energy Agency, Paris. See: www.nea.fr/html/pub/newsletter/2002/20-2-Nuclear_fuel_resources.pdf.

REEP (2000) *Federal Energy Subsidies.* Renewable Energy Policy Project, REEP-CREST, Washington, DC.

Renewable Energy (2004) Federal German government Conference. Issues paper for the International Conference for Renewable Energies, Bonn, June.

Royal Society (1999) *Nuclear Energy: the future climate.* Royal Society/Royal Academy of Engineering, London.

SDC (2006) *The Role of Nuclear Power in a Low Carbon Economy.* Sustainable Development Commission, London, March www.sd-commission.org.uk.

van Leeuwen J.S. and P. Smith (2005) *Nuclear Power: the Energy Balance* http://www.stormsmith.nl/.

WADE (2004) *The WADE Economic Model: China,* a WADE Analysis, World Alliance for Decentralized Energy, Edinburgh, Dec.

WEC (2004) *Comparison of Energy Systems Using Life Cycle Analysis.* World Energy Council, London.

Wind Power Monthly (2006) On-line Focus article and linked Editorial and Feature article 'Nuclear and wind: Serious technical and economic conflicts rule out plans for nuclear and wind mix', *Wind Power Monthly,* June.

WNA (2005) *Critique of 2001 paper by Storm van Leeuwen and Smith: 'Is Nuclear Power Sustainable? and its May 2002 successor: Can Nuclear Power Provide Energy for the Future; would it solve the CO2-emission problem?' with reference to a 2005 version entitled 'Nuclear Power, the Energy balance',* WNA Supplement, (Aug 2002, LCA comparisons updated 2005), World Nuclear Association web site: http:www.world-nuclear.org/.

Part IV Some Key Nuclear Issues

8
New Information on Radiation Health Hazards

Ian Fairlie

Introduction

This chapter discusses new information on radiation health hazards and the extent to which this may impinge on possible decisions concerning new nuclear build in the United Kingdom. It presents a challenging, not to say critical, view of radiation risks, based largely on the findings of the 2004 UK Committee Examining the Radiation Risks of Internal Emitters (CERRIE) report (CERRIE, 2004). It focuses on health risks from radiation exposures rather than the risks of low probability, high consequence, accidents at nuclear facilities, although these can have extremely serious effects as occurred at Chernobyl in 1986. The chapter briefly discusses Chernobyl's effects, the 'dread' factor of radiation and the widely polarised views held on radiation risks. These are partly due to official interest in controlling public perceptions on radiation risks, and the public's apprehension of this.

Difficulties exist in determining the level of effects from radiation at low doses and low dose rates, as explained by the CERRIE report. This found major uncertainties in the internal dose coefficients for radionuclides commonly discharged from nuclear facilities. Although such uncertainties span both up and down from currently accepted central values, the report recommended adopting a precautionary approach.

The chapter indicates that radiation exposures from the current nuclear industry are relatively small, and, if anything, likely to be slightly lowered in any programme of new nuclear build. However the uncertainties in dose coefficients considerably exceed these reductions, and it is concluded that it would be preferable to examine less problematic options for future electricity supplies.

The Chernobyl accident

On 26 April 1986, the world's worst nuclear accident occurred at the Chernobyl nuclear power station in Ukraine which resulted in continuous large releases of dangerous radionuclides into the atmosphere for 10 days (Fairlie and Sumner, 2006). The radioactive fallout from the accident was eventually distributed throughout the northern hemisphere with most deposited in Western Europe. The radiation doses received by liquidators and populations in Belarus, Ukraine and Russia from the fallout were very high. Doses to Western Europe were also very high, resulting in large collective (i.e. population) doses of more than 500,000 person-sieverts. The disaster resulted in about 50 deaths among the emergency workers and liquidators from deterministic (i.e. direct cell killing) effects, but will also result in tens of thousands of cancer deaths from stochastic effects (IAEA/WHO 2005a, IAEA/WHO 2005b).

The precise number of predicted excess cancer deaths from Chernobyl remains a political issue. In September 2005, International Atomic Energy Agency/World Health Organisation issued a Press Release stating that 4,000 extra cancer deaths were expected. However this was inaccurate and was later withdrawn (Edwards, 2006). WHO has now stated on its website that 9,000 excess cancer deaths are anticipated. Recently, scientists from the WHO's International Agency for Research on Cancer (IARC) have estimated that about 16,000 excess deaths will occur (IARC, 2006).

However independent estimates are higher. Fairlie and Sumner (2006) have calculated, using 1988 data from the UN Scientific Committee on the Effects of Atomic Radiation, that between 30,000 and 60,000 excess cancer deaths are expected from doses received from the Chernobyl fallout across the northern hemisphere. These estimates have been supported by others including scientists from IARC (Edwards, 2006).

The effects from the Chernobyl catastrophe in Belarus, Ukraine and Russia are mind-numbing by any yardstick. The health effects include not just cancer deaths, but non-fatal cancers (e.g. most thyroid cancers), non-cancer effects, genetic and teratogenetic effects and psycho-social effects. But there are many other effects including the huge economic consequences, and the fact that very large tracts of Belarus, Ukraine and Russia will be effectively uninhabitable for hundreds of years. Smaller areas will be permanently uninhabitable. Many cancers have latency periods of 20–60 years which means that many cancers are expected to arise in the future: even in 2040 cancer deaths from Chernobyl will still be occurring.

As the real scale of the disaster unfolds, Chernobyl should give us pause for thought before we embark on any revival of nuclear power. Even though future reactors have been stated to be inherently safer than the Chernobyl design, accidents can still occur and it is important that robust plans are agreed internationally for dealing with any future accidents (see Williams, 2001). We should keep in mind the view of the philosopher George Santayana that governments unable to learn from history are condemned to repeat it.

Radiation – a 'dread' issue

Radiation[1] is feared by many people for various reasons. Many observers (see, for example, Slovik, 1987) have remarked that people fear risks that are perceived as

- involuntary,
- inequitably distributed,
- difficult to avoid (i.e. inescapable),
- causing hidden or irreversible damage,
- dangerous to children and future generations,
- causing dread illnesses, i.e. cancer,
- poorly understood by science, and
- the subject of conflicting views by scientists.

Radiation, perhaps uniquely, not only scores on all these perceptions but scores highly, and as a result is usually viewed with unusual degrees of apprehension by the public – see discussion in Meara (2002). It is likely that these fears lie at the foundation of concerns among members of the public about nuclear power and nuclear weapons as observed in surveys and public opinion polls.

Polarised views on radiation risks

Perhaps partly because of this dread factor, widely opposing views exist on radiation risks. Before discussing these, it is useful to state what we do know about radiation's risks. Quite simply, radiation is an accepted carcinogenic, mutagenic and teratogenic agent, even at relatively low levels of exposure. Evidence is beginning to emerge from Chernobyl that radiation exposures may cause non-cancer effects as well, particularly cardiovascular effects (IAEA/WHO, 2005a).

Indeed the low levels of background radiation we all receive, often presented as benign, are instead a killer. The UK Health Protection

Agency – Radiation Protection (formerly the National Radiological Protection Board) has estimated that each year about 6,000 deaths, that is, about 5% of the number of annual UK cancer deaths, are due to background radiation (Robb, 1994). It is true that we cannot identify the individuals who die each year from background radiation, but given the collective dose to the UK's population, it is a relatively simple matter to estimate the number of resulting cancer deaths.

Background radiation is also thought to be the main reason for all childhood leukemias (Baverstock, 2003). It is the reason why women above 40 years have more miscarriages/spontaneous abortions. This is because their ova have been receiving background radiation for 40 or so years – which results in so much damage to their stocks of ova that many, if not most, are not viable. Also, background radiation is intimately connected with the ageing process and is partly why we die from 'old age'. That is to say, radiation is an important contributor to the range of deleterious agents affecting our cells and cellular defence mechanisms. These include radiation, viruses, bacteria, injuries and toxic chemicals. On the other hand, a handful of radiation scientists, whose views lie on the outer fringes of scientific respectability, adhere to the notion that small doses of radiation are actually good for humans (see Jaworowski, 2004). These views on hormesis have been refuted at length by orthodox institutions, see UNSCEAR (1994).

Dose–response relationships at low doses and low dose rates

Given that radiation causes cancers and genetic mutations, the important question is what are the risks at very low levels of exposure. In other words, are there dose levels below which exposures are safe? Currently we think that the answer is no: in other words, no matter how low the radiation dose, there is still a finite (but very low) risk.

In fact it is difficult to establish the precise level of radiation risks at low doses and low dose rates. This is important as most exposures to radiation are at such levels. For example, the overwhelming majority of doses to the public from the nuclear industry are from exposures at very low doses and dose rates. Figure 8.1 indicates the possible dose–response relationships for radiation. Good data exists at high doses (upper right on graph), but almost no data exists at low doses, below 100 mSv, (lower left on graph), so that the various curves shown are all theoretically possible. The simplest and most direct interpretation is the linear (i.e. middle) relationship, and to its credit, the International Commission on

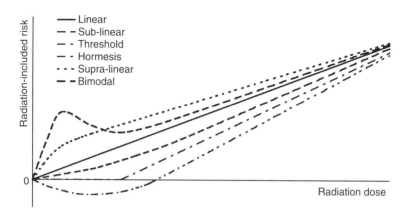

Figure 8.1   Possible dose–responses

Radiological Protection (ICRP)[2] presumes such a relationship for the purpose of setting radiation standards.

Government involvement

A second reason why members of the public are apprehensive about radiation is their perception that major interests are involved when decisions are made on radiation risks. Governments are necessarily involved in policy formation, regulation and standard-setting for both the civil and military nuclear industries. In the United Kingdom, this includes the Nuclear Decommissioning Authority, UKAEA, British Energy, Ministry of Defence, the Naval Dockyards and Rolls-Royce. Although some of these bodies are theoretically in the private sector, in practice they adhere closely to government policies. In addition, the UK Department of Health oversees the regulation of medical practices (including X-ray diagnostic procedures, nuclear medicine and cancer treatment).

These three sectors (civil nuclear, military nuclear and medical) are the main anthropogenic sources of radiation to the public in the United Kingdom. These sectors are fairly powerful establishments in British politics as their views carry considerable weight within the UK government. In addition, the views of the very large regulatory/advisory network governing radiation practices are also very influential. In the United Kingdom, this includes the EA, SEPA, FSA, HSE, NII, HPA–RP (formerly NRPB), UKAEA, DEFRA, DTI, COMARE, DH and MOD.

For these reasons, it is considered that consumer and environmental groups have little influence on official decisions and policies on radiation matters. As Professor G.A. Rose (expert epidemiologist witness to the Black Enquiry) has stated:

> When investigating the environmental health impact of large industries, especially with military interests, we are confronting the seat of immense economic and political power ... (we) constitute no more than an innocent and ill-equipped David confronting Goliath, the well-armed and experienced giant. (Rose, 1991)

This would be less important if decisions on radiation risks were seen to be made strictly according to the scientific evidence available and were not influenced by official policies. Unfortunately, many instances in the past have occurred where industrial and/or governmental policies may have carried much greater weight than scientific considerations. Some technical examples are as follows:

- In 1969, the 70% decrease of tritium's relative biological effectiveness (RBE)[3] by the ICRP contrary to the available scientific evidence (Dunster, 1969).
- In 1980s and 1990s, the ICRP's refusal to recognise Professor Stewart's evidence on radiation risks (Stewart, 1991).
- In 1990, opposition by uranium mining/milling industry to improved radiation safety limits, resulting in laxer limits for workers than had been first proposed.
- In 1990 and 2004, the ICRP's refusal to discuss increased RBEs of beta and Auger emitters.[4]
- In 2004, the ICRP's proposals, in effect, to dilute exposure limits to radiation (see http://www.icrp.org/remissvar/viewcomment.asp?guid={1DAF74F3-05CB-43A7-96CE-E296E0FF061E}.

Also, the history of radiation research is littered with instances of radiation scientists suffering dismissal, blocked careers, loss of research funding and official obloquy because they reported findings on radiation risks which were not welcomed by official agencies or authorities. This has occurred much more often in the US rather than in Europe, but it has occurred in the United Kingdom as well. The following scientists have experienced such discrimination:

- In US – Tamplin, Gofman, Bross, Natarjan, Johnson, Gould, Sternglass, Mancuso, Morgan, Bertell, Radford, and others

- In UK – Professor Alice Stewart and George Kneale
- In Belarus – Professor Y. Bandashevski (imprisoned until recently for publishing his research findings on radiation risks).

The International Commission on Radiological Protection

When discussing official involvement in radiation matters, it is necessary to examine the role of the ICRP, a key establishment in establishing public and worker radiation limits. Its name suggests that it is an official organisation, but in fact it is a voluntary body much like a trade association. However its recommendations are influential within the governments and nuclear industries in many developed nations except the United States. Until recently, the ICRP's 2004 draft recommendations which would have weakened radiation protection standards were effectively withdrawn as a result of an unprecedented level of objections from regulatory bodies, industry representations and environment NGOs (MacLachlan, 2005). A partial list of objectors includes

- Baverstock (2002)
- Schrader-Frechette and Persson (2001)
- The former UK National Radiological Protection Board (see ICRP website http://www.icrp.org/remissvar/viewcomment.asp)
- Proceedings of the EC Stakeholders' Conference on environmental radioactivity, December 2002, in Luxembourg, see http://europa.eu. int/comm/energy/nuclear/radioprotection/doc/conference/shc_ 2003_09_19_proceedings_en.pdf
- The rejection of certain provisions in the ICRP draft by the IAEA's Safety Standard Series Guidance on Exclusion, Exemption and Clearance levels
- The comments of the UK National Dose Assessment Working Group (see ICRP website) http://www.icrp.org/remissvar/listcomments.asp
- The EC's Article 31 Group of Experts, at their meeting on the ICRP proposals in Luxembourg in November 2004, is understood to have criticised many proposals in the draft, and recommended further time for the draft's consideration or for its complete revision
- Particularly trenchant objections were lodged at the ICRP website by the environment NGO organisation, Greenpeace International – see http://www.icrp.org/remissvar/viewcomment.asp?guid={07A7B32B-137B-4009-8F2445A21B36B98B}
- Other unflattering comments from members of the public on the ICRP's (now withdrawn) draft recommendations are available for perusal at http://www.icrp.org/remissvar/viewcomment.asp.

Such objections to ICRP recommendations are not a new phenomenon. In the past, many authors have criticised the ICRP for downplaying the hazards of radiation and for its lax recommendations on radiation risks, including Caufield (1990), Greenberg (1991), Rose (1991), Stewart (1991), Proctor (1995), Greene (1999) and Baverstock (2002, 2005). Rather depressingly, these and other criticisms have met with little effect.

The CERRIE report

With the presence of such conflicting views on radiation risks, where can we turn to for a balanced yet informed discussion on radiation risks? The 2004 report of the CERRIE is a useful start. The CERRIE committee was established in 2001 by the UK government to look into internal radiation risks following concerns about the effects of ingested or inhaled radio-nuclides. Implicit in the committee's terms of reference was the need to examine a number of hypotheses which suggested that the effects of low levels of radiation were much greater than currently recognised by official bodies including the ICRP and the UK's National Radiological Protection Board (now subsumed within the Health Protection Agency and retitled HPA–RP: Health Protection Agency–Radiation Protection). The committee's membership reflected the wide spectrum of views on radiation risks, as it included representatives from environment groups, NRPB, BNFL, and independent, critically minded scientists. Its report was published in October 2004. Committee Examining the Radiation Risks of Internal Emitters (CERRIE) is perhaps the first balanced committee established by the UK government (i.e. with representatives having critical views about radiation risks) to look into radiation risks. Because of its balanced membership, the CERRIE report rewards close reading. It gives short shrift both to theories that small doses of radiation are good for you and to theories that it may be thousands of times more hazardous than currently acknowledged. Unlike many official reports on radiation matters which are usually incomprehensible, the CERRIE report was written to be understandable by lay members of the public. Although the report indicates matters on which its members could not agree, the arguments are set out for the public (and professionals) to read for themselves.

The report's key virtue is that it discusses issues which radiation authorities would perhaps rather have not publicised. These include the large uncertainties involved in official estimates of internal radiation doses; the possible dangers of radionuclides emitted from nuclear facilities and of commonly administered nuclear medicines; and the newly discovered strange effects of radiation which question previously accepted theories about how radiation exerts its harmful effects, that is, through ionisa-tion damage on DNA. Radiation may still adversely affect DNA, but this is not its only effect.

The report's main conclusion was that internal radiation doses, that is, from ingested or inhaled nuclides, are associated with substantial uncer-tainties. The main reasons were

1. uncertainties in the models used to estimate the internal uptake, reten-tion and excretion of ingested or inhaled nuclides and the distribu-tion of these nuclides in the body and cells;
2. natural variability between humans; and
3. uncertainties about the derivation of dose coefficients.

Table 8.1 indicates uncertainties in the dose coefficients for some nuclides spanning several orders of magnitude which were considered by the Committee. The report concluded that such uncertainties required government policymakers and regulators to adopt a precautionary approach when considering exposures to internal radiation.

The CERRIE report also considered newly discovered effects of radiation, including genomic instability (ongoing, long-term increase in mutations within cells and their offspring), bystander effects (adverse effects in unir-radiated cells next to cells which are irradiated) and minisatellite muta-tions (inherited germ-line DNA changes). It stated that these were real biological events which could well have a significant impact on radiation risks. These new effects needed further research.

Table 8.1 Uncertainties in dose coefficients

Nuclide	Intake method	Organ	Range = (95th/5th percentiles)
Cs-137	Ingestion	Red bone marrow	4
I-131	Inhalation	Thyroid	9
Sr-90	Ingestion	Red bone marrow	240
Sr-90	Ingestion	Bone surface	390
Pu-239	Ingestion	Red bone marrow	1,300
Sr-90	Inhalation	Lungs	5,300
Ce-144	Inhalation	Red bone marrow	8,500
Pu-239	Ingestion	Bone surface	20,000

Source: Goossens, L.H.J., F.T. Harper, J.D. Harrison, S.C. Hora, B.C.P. Kraan and R.M. Cooke (1998) *Probabilistic Accident Consequence Uncertainty Analysis: Uncertainty Assessment for Internal Dosimetry: Main Report*. Prepared for US Nuclear Regulatory Commission, 20555-0001, Washington, DC, USA and for Commission of the European Communities, DG XII and XI, B-I049, Brussels, Belgium.

The report stated that uncertainties in dose coefficients for some internal radionuclides (e.g. plutonium-239) could be very large. Table 8.1 indicates the ratios between the 95th and 5th percentiles of the probability distributions for organ dose coefficients (i.e. Sv per Bq) of various nuclides. It can be seen that the ranges of some nuclides can be very large indeed. These uncertainties operate both up and down, that is, both to increase or decrease possible risks.

However it is clear that more attention should be devoted to the former than the latter possibility (i.e. that the risks may be greater) because

a. the Precautionary Principle (Hey, 1995) requires us to choose the option which results in less damage in case we get our risk assessments wrong, i.e. we need to be concerned with the possibility of greater risks;
b. the new biological evidence, especially of bystander effects, suggests that risks at low doses may be greater than those estimated by a linear extrapolation from high doses; and
c. evidence exists (the leukaemia clusters near Sellafield and other nuclear sites) which suggests that the radiation risks of some radionuclides may be greater than currently acknowledged.

In addition, ever since radioactivity and radiation were discovered over a century ago, our understanding of their risks has always increased: public and worker dose limits have always been tightened.[5] It would be unwise to presume that this process has stopped. Therefore when considering whether radiation risks could be higher or lower by factors of <10, it is prudent to pay greater attention to the former, than the latter, risks. Of course, the nuclear industry tends to the opposite conclusion, i.e. that the risks might be smaller and therefore there is no need to devote resources to reducing them, especially in a competitive electricity market. Ultimately, the question as to what we should do faced with these uncertainties is a political one, although the CERRIE report clearly recommends that a precautionary approach be adopted.

Practical implications of the CERRIE report

The new scientific evidence in the CERRIE report poses challenges to regulators and policymakers in radiation protection. Because the Committee's terms of reference were restricted to examining the scientific evidence, the Report does not contain policy recommendations apart from future scientific research. However the government has maintained on many occasions that its policies on radiation protection are based on the latest available scientific evidence. The significance of the CERRIE report is that it presents new scientific evidence that will eventually be required to be reflected in future policies, guidance and regulatory practices. Most important is the large uncertainty surrounding internal radiation doses and their risks. In future, the Report recommends that the Precautionary Principle will need to be used when dealing with internal radiation exposures, and it is likely that new regulatory provisions will be needed. Indeed, the first sentence of the Report's press release states: 'Tougher action is needed to allow for new information about the risks from internal radiation.'

Effects on the nuclear industry

A well-orchestrated campaign (Leake, 2005) has pressed the case for the construction of new nuclear power stations in the United Kingdom. One of the issues that has to be addressed is that of radiation exposure. Radiation doses to the public from the UK nuclear industry are relatively small. For example, doses from medical diagnoses and treatments are much larger, though such exposures are voluntary and have a countervailing medical benefit. The nuclear industry is fond of quoting the 'statistic' that the average radiation dose to a member of the UK public from its activities is very small. This may be theoretically correct but such exposures have been divided by the UK population of ~60 million. The 'statistic' is therefore irrelevant at best, and at worst, misleading, as exposures from the UK nuclear industry are highly localised and mainly affect nuclear workers and the relatively small populations of those living near to, or downwind from, nuclear facilities.

The largest UK *internal* radiation exposures are from the nuclide discharges to air and sea from the Sellafield reprocessing plant in Cumbria. The normal operation of the nine remaining UK nuclear power stations results in much smaller internal radiation doses to the public. *External* radiation exposures largely concern workers, with Sellafield again being the largest contributor, with smaller doses at nuclear stations. Less uncertainty surrounds the estimation of external radiation doses.

Any proposed expansion of nuclear electricity generation would inevitably entail some additional exposure to radiation. In the absence of detailed plans, it is difficult to state whether such doses would be similar to, greater or less than, those currently experienced. An important factor would be the decision whether to reprocess spent nuclear fuel from new proposed nuclear stations. This has not been indicated one way or the other in official statements, but informal indications are that the UK government has been made aware of the manifest technical, safety, and economic blunders of reprocessing and that no further reprocessing is being proposed. If so, net public and worker doses from new nuclear facilities would be lower than from present facilities. Radiation doses would continue to occur, albeit probably at lower levels.

However the main factor is not the likely low level of exposures from any new nuclear facilities, but possible increased perceived risks from radiation doses, due to our better understanding of radiation's effects. It might be possible, through informed policy choices, to reduce population and worker doses from new nuclear stations by factors of two or three. But the uncertainties surrounding internal radiation are larger, and in the case of some radionuclides, much larger.

It is concluded that the answer to the question – does the latest scientific evidence pose questions for new nuclear build – would appear to be a qualified yes. Given the latest evidence, it would be preferable to examine less problematic options for future electricity supplies.

Notes

1. This chapter is concerned with ionising radiation, i.e. radiation with sufficient energy to ionise atoms, which includes alpha particles, beta particles, gamma rays and neutrons. It excludes non-ionising radiation, such as infrared radiation and ultraviolet radiation.
2. The International Commission on Radiological Protection (ICRP) is a voluntary body of radiation scientists whose recommendations on radiation protection have considerable influence in most countries.
3. Relative biological effectiveness (RBE) is a measure of the damaging nature of a particular form of radiation.
4. Auger emitters are nuclides which decay by means of dense showers of very low-range electrons.
5. In 1934, the occupational limit for radiation was equivalent to ~1.2 mSv (millisieverts or one-thousandth of a sievert) per day. This was tightened in 1951 to 3 mSv per week, in 1966 to 50 mSv per year, and in 1990 to the present limit of 20 mSv per year averaged over five years (with a maximum of 50 mSv in any one year) (from Stather, 1993). Expressed in per annum terms, the limits were 438, 156, 50 and 20 mSv per year.

References

Baverstock, K. (2002) Letter, *J. Radiol. Prot.* **22** (December), 423–424.
Baverstock, K. (2003) 'Childhood leukemias are caused by background radiation', *New Scientist*, 9 January 2003, p. 4.
Baverstock, K. (2005) Science politics and ethics in the low dose debate. *Medicine Conflict and Survival* **21** (2), 88–100.

Caufield, C. (1990) *Multiple Exposures: chronicles of the radiation age*. Penguin Books, London.

CERRIE (2004) *Report of the Committee Examining Radiation Risks of Internal Emitters*. www.cerrie.org.

Dunster, H. (1969) Progress Report from the ICRP. *Health Physics* **17**, 389–396.

Edwards, R. (2006) 'How many more lives will Chernobyl claim?' *New Scientist*, 6 April 2006, p.11. http://www.newscientist.com/article/mg19025464.400-how-many-more-lives-will-chernobyl-claim.html.

Fairlie, I. and D. Sumner (2006) *The Other Report on Chernobyl* (TORCH), Published by Greens/EFA Group of the European Parliament. http://www.greensefa.org/cms/topics/dokbin/118/118499.the_other_report_on_chernobyl_torch@en.pdf.

Greenberg, M. (1991) The evolution of attitudes to the human hazards of ionising radiation and to its investigators. *Am J. Ind. Med.* **20**, 717–721.

Greene, G. (1999) *The Woman Who Knew Too Much*. University of Michigan Press, Ann Arbor, MI, USA.

Hey, E. (1995) *The Precautionary Principle. Where Does It Come From And Where Might It Lead In The Case Of Radioactive Releases To The Environment*. In: Proceedings of an International Atomic Energy Agency Symposium on The Environmental Impact of Radioactive Releases. Vienna, May 1995. IAEA-SM-339/195.

IAEA/WHO (2005a) *Health Effects of the Chernobyl Accident and Special Health Care Programmes*. Report of the UN Chernobyl Forum Expert Group "Health" (EGH) Working draft. 26 July 2005.

IAEA/WHO (2005b) *Environmental Consequences of the Chernobyl Accident and their Remediation*. Report of the UN Chernobyl Forum Expert Group "Environment" (EGE) Working draft. August 2005.

IARC (2006) *The Cancer Burden from Chernobyl in Europe*, International Agency for Research on Cancer, World Heath Organisation, Paris. http://www.iarc.fr/ENG/Press_Releases/pr168a.html (http://www3.interscience.wiley.com/cgi-bin/abstract/112595693/ABSTRACT).

Jaworowski, Z. (2004) Chernobyl, nuclear wastes and nature. *Energy and Environment* **15/5** (October), 807–824.

Leake, J. (2005) 'The nuclear charm offensive', *New Statesman*, 19 May 2005.

MacLachlan, A. (2005) *Nucleonics Week*. Volume 46, Issue 28, 14 July 2005.

Meara, J. (2002) Getting the message across: Is communicating the risk worth it? *J. Radiation Protection* **22**, 79–85.

Proctor, R.N. (1995) *Cancer Wars: how politics shapes what we know and don't know about radiation*. Basic Books. New York, NY, USA.

Robb, J.D. (1994) *Estimates of Radiation Detriment in a UK Population*. NRPB Report R-260, National Radiological Protection Board, Chilton Oxon.

Rose, G.A. (1991) Environmental health: Problems and prospects. *J. Royal College of Physicians of London* 25 (1), 48–52.

Schrader-Frechette, K. and Persson, L. (2001) Ethical, logical and scientific problems with the new ICRP proposals. *J. Radiol. Prot.* **22**, 149–161.

Slovik, P. (1987) Perception of risk. *Science* **236**, 280–285.

Stather, J.W. (1993) *Radiation Carcinogenesis – Past, Present and Future*. In: Proceedings of an NEA Workshop. Radiation Protection on the Threshold of the 21st Century, Paris, January 1993. NEA/OECD, Paris, pp. 21–37.

Stewart, A.M. (1991) Evaluation of delayed effects of ionising radiation: An historical perspective. *Am. J. Ind. Med.* **20**, 805–810.

UNSCEAR (1994) *Sources and Effects of Ionising Radiation*. United Nations Scientific Committee on the Effects of Atomic Radiation, Vienna.

Williams, D. (2001) The world needs to improve its handling of international disasters (editorial). *BMJ* **323**, 643–644. 28 March 2001.

replaced the first appointee as CoRWM chair, who resigned almost immediately to head Northern Ireland's water quango.

Interviewed in the *Guardian* newspaper (27 September 2005) Professor MacKerron opined: 'A lot of my history has been in public-domain debating. I always knew, when I took on the chairmanship, that it would be controversial.'

Controversy has regularly reared its head around CoRWM, led by disgruntled peers in the House of Lords, whose select committee on science and technology expressed their deep concern at slow progress towards developing policy on radioactive waste management. MacKerron swats away such moaning: 'There's a certain amount of frustration that we were expected to start with a blank sheet of paper. Members haven't fully appreciated that we are about process as much as substance. We have to have a properly audited trail.'

Perhaps inevitably, the search for a solution to the safe management of nuclear waste became mixed-up with the emergent energy policy debate, in particular, the arguments for a nuclear renaissance of new-build reactors. But ministers seem to have misunderstood – or possibly deliberately obfuscated – the reason why they created CoRWM. For example, Energy Minister Malcolm Wicks told parliament in November 2005: 'Options for the long-term management of higher activity wastes in the UK are currently the subject of consideration and evaluation by CoRWM, who are due to make their final recommendations to the Government in 2006. The long-term management policy for these higher activity wastes will then be decided by the UK Government and the devolved administrations in the light of CoRWM's recommendations.'[5]

However, he later elaborated stating: 'CoRWM has already considered some new build scenarios, drawing technical information from industry sources. The results of this work were published in the CoRWM Inventory in July 2005. CoRWM have confirmed that waste from a new build programme could be technically accommodated within any of the options they have short-listed for long-term waste management.'[6]

But as Professor MacKerron, in his position as chair of CoRWM, strongly emphasised in a letter to the *Independent* newspaper early in 2006, and repeated in evidence to the Trade and Industry Select Committee (TISC) on 19 June 2006,[7] 'as a committee, we have no position on the desirability of nuclear new-build. Our recommendations should not be seen as either a red or green light for new reactors. It is not our place to set a timeframe for Government decisions on new-build, although we do believe they should be subject to their own assessment process, including the consideration of waste. This is because such decisions raise different political and ethical issues when compared with the consideration of wastes that already exist.'

Several months later, the influential *New Scientist* magazine (6 May) editorialised: 'Some advocates of nuclear power will doubtless argue that CoRWM has now provided that plan ... This is optimism gone mad. Deciding to put waste down a hole, with no idea what form the repository should take or where it should be, is no more of a plan than has existed for the past 30 years.' There is ample evidence to suggest that is a correct observation, as will be seen later.

CoRWM's roller-coaster ride in framing public policy on radioactive waste has involved dozens of meetings, uniquely for a quango, almost all sessions taking place in public; as the debate swirled around it, often adding new issues for consideration.

For example, a week after London was awarded the Olympic games in July 2005, it was revealed by a Conservative London Assembly member that part of the planned Olympic Park is on the site of a small former research nuclear reactor – decommissioned in 1982 – and the land around it may be radioactively contaminated.[8] But after the nuclear safety regulator, the Nuclear Installations Inspectorate, gave a clean bill of health, the story died a death.

Eroding confidence

Another issue that raised its head with which CoRWM had to grapple was the potential problem of the threat of inundation and erosion of prospective disposal/long-term storage licensed nuclear sites. One of the main reasons the Prime Minister Tony Blair cited in justifying bring nuclear power back onto the agenda 'with a vengeance' was his clear concern over threats from climate change.[9] His chief scientific advisor, Professor Sir David King, has played key role in convincing Mr Blair not only of the catastrophic impact of uncontrolled climate change, but also of the atomic option as significant part of the solution.

It is thus ironic that Mr Blair was reported as backing the building of a new fleet of reactors at currently licensed nuclear sites, as they have been identified as vulnerable to the very sea-level rise and coastal erosion Professor King had championed in his warnings on the importance of adaptation to the exigencies of climate change.

In December 2005, UK Nirex, the Nuclear Industry Radioactive Waste Executive, Britains nuclear waste management agency (which is now quasi-independent, but had previously been part of the industry), published a report as part of their input into CoRWM's environmental evaluation of the UK's nuclear waste management options over the next 300 years.

Nirex's summary of 'Climate and Landscape Change' at nuclear sites operated by the newly created (April 2005) nuclear quango, the Nuclear Decommissioning Authority (NDA) sites, was as follows (Table 9.1): [One other reactor site, Torness in Scotland, owned by British Energy is not covered.]

The Nirex study makes use of an ongoing project, Future Coast,[10] conducted by the Defra. Indeed, the then Defra environment minister Elliot Morley, sacked in a ministerial reshuffle in May 2006 (it seems for his nuclear-sceptic leanings), said in an answer in Parliament[11]: 'Operational responsibility for managing the risk from coastal erosion in England rests with maritime district councils who, in partnership with the Environment Agency and other bodies with coastal defence responsibilities, take an integrated and long-term view of managing coastlines through shoreline management plans. These plans, in line with government policy, consider the implications of coastal processes, including erosion. More detailed coastal strategies are then developed, taking into account economic, social and environmental matters. These detailed strategies consider the specific

Table 9.1 Vulnerability of NDA sites to landscape change

Site	To 2020	To 2100
Sizewell	Possible	Vulnerable to erosion[a]
Bradwell	Unlikely	Vulnerable to inundation[b]
Dungeness	Possible	Very vulnerable to erosion[c]
Winfrith	Very unlikely	Unlikely[d]
Hinkley Point	Unlikely	Possible[f]
Berkeley/Oldbury	Unlikely	Vulnerable to inundation[g]
Wylfa	Unlikely	Unlikely[h]
Chapelcross	Very unlikely	Very unlikely[e]
Sellafield	Unlikely	Vulnerable to erosion[i]
Hunterston	Unlikely	Unlikely[h]
Dounreay	Unlikely	Unlikely[h] – possible for the shaft

[a] Vulnerable to loss of sediment from the North.
[b] Vulnerable to subsidence, rising sea level and rollover of the Blackwater estuary.
[c] Very vulnerable to change in sediment availability, drift direction and human intervention.
[d] Low risk of disruption by fluvial processes.
[e] Inland site with very low rate of change (unverified).
[f] Present massive sea defences provide protection.
[g] Progressive marine transgression likely to claim the sites unless protected.
[h] Hard rock coastal headland with low rate of erosion (unverified).
[i] Currently considered to be protected by swash alignment. Vulnerable to change in wave climate. Irt estuary likely to flip with rising sea level.

needs of key coastal installations, such as nuclear power stations, Sellafield and the low-level waste repository (LLWR) at Drigg.'

He added: 'The impact of coastal erosion on the LLWR has also been considered as part of a Post-Closure Safety Case for the site, submitted to the Environment Agency by British Nuclear Group Sellafield Limited (BNGSL) in September 2002. BNGSL's assessment predicts that the LLWR could be destroyed by coastal erosion in 500 to 5000 years if no action were taken to maintain the coastline.'

Former prospective disposal sites revealed

In June 2005, following several freedom of information requests, the long-kept-secret list of prospective disposal sites for nuclear waste , known to have been drawn up by Nirex, was finally released. It contained a list of 537 sites in all, but boiled it down to 12 key locations[12]:

Bradwell & Potton Island, Essex; two sites at Sellafield, Cumbria; Dounreay & Altnabreac, Caithness; Fuday & Sandray, Hebrides; Killingholme, South Humberside; Stanford, Norfolk; Offshore site near Redcar, Northumberland; and Offshore site near Hunterston, Scotland.

Nirex managing director Chris Murray said: 'We hope that the publication of the list, following consultation with our stakeholders, will help to move the debate away from past attempts to tackle this issue and on to the new process, led by CoRWM, in which we would encourage everyone to get involved.'[13]

But despite ministers and Nirex denying the sites were still being actively considered, the revelations resulted an a flurry of local media and public concern, fearing their community might again be earmarked as a future nuclear waste disposal site.[14]

Volume confusion

Another issue that continued to return to the debate was the volume of radioactive waste for which CoRWM was planning its management strategy. Mr Morely told the then Conservative energy spokesman Bernard Jenkin in a written reply[15] in October 2005 that the UK Radioactive Waste Inventory –as jointly prepared by Defra and Nirex – gives details of the volumes of waste in store in the United Kingdom as on 1 April 2001 (Table 9.2):

Mr Morley also provided[16] the then Liberal Democrat front bench environment spokesman, Norman Baker, with projected future volumes of radioactive waste arisings – assuming projected closure of nuclear power stations occurs – as given in Table 9.3.

Table 9.2   Volumes of waste in store in the United Kingdom as on 1 April 2001

Types of Waste	Volume (Cubic metres)
Higher-level waste	764
Intermediate-level waste	74,466
Low-level waste	15,674

Notes: HLW: higher-level waste; ILW: intermediate-level waste; LLW: low-level waste.

Table 9.3   Projected future volumes of higher activity radioactive waste

Date	High level (cubic metres)	Intermediate level (cubic metres)
2010	1,350	107,000
2020	1,510	128,000
2030	1,510	143,000

The striking detail not presented in these figures is the 20,000,000 cubic metres of projected radioactively contaminated land estimated by the NDA in its draft national strategy – issued in August 2005 – to exist at Sellafield alone. As the Strategy put it bluntly: 'At Sellafield, it has been estimated that there may be as many as 20 million cubic metres of contaminated land, caused mainly by leaks from legacy and disposal facilities.'[17]

Military legacy

Strictly speaking, CoRWM was not asked to include in its mission proposals to deal with the significant quantities of radioactive waste that has arisen from Britain's military nuclear programme, dating back to the late 1940s. An indication of the immediate costs of the care and maintenance handling – not long term management – of some of these waste streams was given by the Ministry of Defence in an answer to parliament in May 2006 (Table 9.4).[18]

Other concerns over military-origin radioactive pollution were raised in respect of contamination of beaches.[19] The *Sunday Herald* newspaper reported in November 2005 that 'in a move which has frustrated the Scottish Environment Protection Agency (Sepa), angered experts and infuriated local residents, the Ministry of Defence (MoD) is refusing to take responsibility for cleaning it up'.

The paper reported that a survey commissioned by Sepa has uncovered nearly 100 radiation hotspots around the shore at Dalgety Bay in Fife.

Table 9.4   Immediate cost of care and maintenance

Financial years	Cost (£ million)
2000–2001	213
2001–2002	201
2002–2003	241
2003–2004	188
2004–2005	253

It added that radioactive contamination up to 48 times higher than normal levels was found at 97 separate locations on the foreshore, according to a report released by Scotland's environmental regulator Sepa.

Scotland's sensitive situation

Indeed, the circumstances of Scotland's nuclear waste burden are particular to that nation, and it has provoked a continuing public debate, as well as resurrecting concern periodically in the Scottish Parliament. The most controversial issue has been the unresolved – and expensive – difficulty of dealing with the decommissioning and radioactive remediation of the so-called Dounreay shaft, an access tunnel backfilled with radioactive detritus from laboratory experiments and other research activities at Dounreay since the late 1950s, and which suffered a major chemical explosion in 1977, leading to considerable localised radioactive contamination of parts of the site and nearby beaches.[20] Local residents in Buldoo, a small village near the plant, have also expressed opposition to a low-level nuclear disposal facility – they see it as a 'dump' – being built on their doorsteps. Concern was further raised when an on-site radioactive waste store was discovered to be leaking early in 2006. A press spokesman for Dounreay – which holds nuclear waste from Belgium, Denmark, France, Germany, Holland and Georgia on site – insisted the level of radioactivity in the surrounding loop was a million times lower than in the silo, saying: 'The measures now in place provide additional reassurance about the safe containment of the wastes, pending its retrieval.'[21]

The United Kingdom Atomic Energy Agency (UKAEA) said the proposed inland site for the facility was chosen to meet concerns about the future effects of global warming on the sea level surrounding the coastal plant. If everything is approved it is expected that work would start early in 2008 and that the facility would be operational by 2011.[22]

Scotland's Labour First Minister Jack McConnell also confirmed that it was Scottish Executive policy that a solution to the problem of waste must come before any decision to build new nuclear plants. 'If it is so safe perhaps the UK Government could dig a hole in the middle of London and store its nuclear waste there', caustically commented Scottish National Party MSP Richard Lochhead.[23]

Nuclear nemesis

Early in January 2006, the *Guardian* newspaper reported[24] 'Ministers warned of huge rise in nuclear waste'; and that 'lethal radioactivity [of nuclear waste] could rise five-fold'. It was an important article, as it debunked an often repeated claim – by ministers and nuclear proponents alike – that any new nuclear power programme would produce only 10% of the amount of radioactive waste that has been generated by the current fleet of British reactors. It is true there would be less overall *volume* since reprocessing was not envisaged with the new programme, so there would be much less ILW and LLW. But there would be a lot more HLW, i.e. the unreprocessed spent fuel.

As a result an analysis undertaken by CoRWM suggested that the spent uranium fuel rods from new power stations would almost triple the radioactivity in the current inventory of UK radioactive waste.

Chris Murray, chief executive of Nirex, was quoted as commenting: 'The volume is not the whole story. We need to be very exact about what type of waste new reactors would actually produce and how it needs to be dealt with.' Assuming Britain would build 10 new reactors – and importantly will not reprocess the spent fuel – Corwm suggested that — such a programme would produce an extra 31,900 cubic metres of spent fuel, on top of the 8150 cubic metres currently stored. More cautiously, Professor MacKerron observed 'The footprint of any facility you might want would have to be increased, by more than 10% but nothing like as much as 2–3 times. It's very difficult to know at the moment where between those extremes it lies.'

Contemporaneously, the scientific weekly, *New Scientist*, reported[25] that the main environmental regulator for England and Wales, the Environment Agency, was expressing concern that the proposed containers for burying nuclear waste could crack and leak within 500 years, making the plans for a deep underground repository 'overly optimistic', according to internal documents obtained by the magazine. Nirex had proposed that radioactive waste from civil and military nuclear programmes be buried between 300 and 1000 metres below ground at a site yet to be selected, but the Nirex report sent to the Environment Agency – both the report and the response from the regulator were posted on their respective web sites contemporaneously – warned that there would be a long-term risk that the concrete and steel waste containers would corrode and fail.

These revelations followed on shortly after another newspaper had revealed that the projected cost of cleaning up the sites of Britain's old nuclear power stations was likely to leap to more than £70 billion – up £14 billion – from the NDA's earlier estimate in 2005. The newspaper pointed out that the new total figure was the equivalent of a charge of £800 for every person – adult and child – in the country.[26]

The Royal Society (RS), effectively the UK's national academy of sciences, chose this moment to attack the composition of CoRWM. In a bitter critique[27] – released on 9 January – based on a scientific seminar held the previous November, the RS said it was 'vital that CoRWM obtains stronger scientific input as it moves into the final stages of its work in reviewing options for managing the UK's radioactive waste'.

The report recommended that scientific and technical organisations should be involved with the exercise to assess the 'weight' that should be given to different criteria being applied to CoRWM's short list of options for the disposal of radioactive waste.

Professor Geoffrey Boulton, co-ordinator of the RS's report and independent member of the CoRWM Quality Assurance Group, said: 'CoRWM has vital role to play in pointing a way forward for the serious and urgent issue of disposal of nuclear waste, and it is the Royal Society's intention to offer constructive advice to aid this important task. We are concerned that the hitherto relatively limited engagement with the scientific and engineering communities, apart from in small specialist groups, might result in a negative response to the final CoRWM proposals. We suggest the Committee seeks to avoid this by engaging now with the scientific and engineering learned societies to complement the public engagement work of CoRWM.'

He added that the RS 'support[ed] the crucial importance of the public consultation and engagement processes that are being managed by CoRWM. It is important that when CoRWM reports, it is credibly able to claim broad public support for the preferred options. Without this, the CoRWM process will have been yet another ineffectual stage in the history of the UK's failure to develop policy for this vital issue.'

The RS report also suggested that Defra should put in place an independent successor to CoRWM because the timescale for a final report in July 2006 was, in its judgment, 'far too short to move from a series of discrete, favoured options' to an integrated strategy based on those options. Such a body will need much greater scientific and technical capacity than CoRWM since accessing the knowledge of the science community and developing a consensus within it will be important in establishing a credible strategy.

The outputs of one of the CoRWM-sponsored expert workshops aimed at 'scoring' comparative hazards, that had worked up the ire of the RS, concerned the issue of nuclear security. The security experts' workshop recommendations warned that Britain's nuclear waste was – and indeed is – vulnerable to terrorist attack and the government was failing to address the issue with sufficient urgency.

The nuclear security specialists – including the author of this chapter – urged the government 'to take the required action and to instruct the NDA, in cooperation with the regulators, to produce an implementation plan for categorising and reducing the vulnerability of the UK's inventory of radioactive waste to potential acts of terrorism, through conditioning and placement in storage options with an engineered capability specifically designed to resist a major terrorist attack'.

The security issues associated with nuclear power, including those associated with transporting nuclear materials, are explored in detail in the next chapter.

Radioactive waste inventory illustrates size of problem

Early in 2006 Nirex published an updated inventory of radioactive waste already stockpiled in the UK, or forecast to be created from planned nuclear operations. The media began to characterise the quanties – some 2.3 million cubic metres – stored around the country – as more than enough to fill the Albert Hall five times. At that time, only 8% of the existing stockpile of radioactive waste material had been securely packaged. Figures based on stocks as of April 2004, showed a 11% decrease in high-level waste – from 1510 cubic metres – since the 2001 inventory. There was also a 7% fall – from 237,000 cubic metres in 2001 – in ILW. As of the beginning of 2006, the NDA estimated that 750 tonnes of overseas spent fuel was being stored at facilities in the United Kingdom, mostly at Sellafield. Ministers still insist that precise details of deliveries from individual customers are 'commercially confidential'. But there was also a significant 35% increase in the mildly radioactive LLW – from 1.51 million cubic metres to 2.1 million, due to recent declarations of suspect contaminated land, according to Nirex.[28] For example, of the additional 470,000 cubic metres of LLW, 370,000 cubic metres is made up of contaminated ground from the Aldermaston Atomic Weapons Establishment (AWE) with a further 20,000 cubic metres of contaminated soil from other sites.

The energy review – more waste?

As the government's energy review – seen by many as a essentially a nuclear review with a pre-determined pro-nuclear conclusion – was launched at the end of January. Professor MacKerron observed that talk of building new nuclear power stations before publication of CoRWM's final report then could undermine the process. 'People expect the waste issue to be resolved before any decision is taken on building new reactors. That was what we had been led to believe was the Government's position', he said, adding 'The Government always made a commitment that it will need to solve the waste problem before a rebuild decision. Given that the report on rebuild is expected in early summer it puts pressure on us.' And Malcolm Wicks described the failure to find a permanent solution to the problem as a 'national disgrace'.[29]

Another Department of Trade and Industry (DTI) minister Lord Sainsbury, responsible for science, somewhat disingenuously reassured peers in a written reply that 'The Government have made it clear that before we can contemplate a new generation of nuclear reactors, we must demonstrate to the public that the legacy' of nuclear waste is being tackled. Under the managing radioactive waste safely (MRWS) programme, there is a clear strategy in place, and work is under way to tackle that legacy.[30]

More believeably, Wicks told increasingly concerned MPs at DTI question time in mid February that 'the clean-up of our waste legacy is one of the big challenges that we face; in my judgment, it should have been tackled before now. We now have in place the NDA, and an expert committee will be advising us in the summer on the equally important issue of a final repository for nuclear waste. Once both of those are in place, we shall be in a position to discuss with the public – should we need to, and should that be our decision – the future of civil nuclear power in this country', adding 'By any judgment, the cost of clearing up the nuclear legacy, calculated over time – possibly 50 or more years – is … a very expensive project'.[31]

Another influential figure, Sir John Harman, chairman of the main environmental regulator, the Environment Agency, told the *Observer*: 'An actual nuclear waste facility is probably 15 years in the future. If a decision was postponed on this, we would think it imprudent to start a new programme of building nuclear reactors not knowing what we are doing about the waste.'[32]

Dump it 'down-under'?

Without prior notice, suddenly a completely new option for the long-term stewardship of radioactive waste was injected into the debate – from the other side of the world, in Australia. Storing nuclear waste in the geologically stable Australian outback was the only international solution to ensure the safety of both Australia and the world, argued nuclear physicist Dr Geoff Hudson, of the University of Melbourne, adding there was no sound reason for Australia 'not to do the world a favour'.[33]

But nothing concrete came of this speculation, so Professor MacKerron was left to options in his original suite. He told the *Sunday Times*,[34] 'What to do with our nuclear waste is a national problem that has not been solved over a long period', pointing out that his main dilemma was choosing between the improved storage of radioactive material, which assumes that Britain will still be politically stable 100 years from now, and an early commitment to deep underground disposal, which means the waste would be out of reach of any future technological advances, adding: 'There is very likely to be some mixture of options in our recommendations. It would be very surprising if one size fitted all from now on.'

CoRWM gets to grips with holistic assessment

Throughout the first half of 2006, CoRWM's round of meetings continued, with virtually all proceedings being held in public, and with details posted promptly on its web site, along with dozens of support documents and papers written by and for CoRWM members. CoRWM's own summary[35] of the event stressed it had 'completed an important part of its assessment of options for the long-term management of radioactive wastes … The purpose of using MCDA [Multi-Criteria Decision Analysis] was to deepen understanding of the relative strengths and weaknesses of the different options, rather than to produce a "right answer" in terms of a best option.'

At its final decision-making plenary meeting on 25–27 April 2006, CoRWM issued its draft recommendations.[36] Their key conclusions and guidance included the decision that geologic emplacement in some form of subterranean repository was the best option. This was qualified with the observation that 'CoRWM recognises that there are social and ethical concerns that might mean there is not sufficient agreement to implement geological disposal at the present time. In any event, the process of implementation will take several decades.' CoRWM therefore recommended a 'staged process of implementation', that inter alia

would involve 'reviewing and ensuring security, particularly against terrorist attacks'.

Political reaction

Political reaction was swift; responding to CoRWM's draft recommendations Shadow Trade & Industry Secretary, Alan Duncan MP said:

> It is vitally important that we have a credible long-term solution to the
> problem of nuclear waste and I welcome the work that CoRWM has
> done on this important issue. We urgently need to address the problem
> of legacy waste anyway, and any new build would require a credible
> waste management policy. Without a long-term regime for waste no
> company would ever invest in a new nuclear power station. Whatever
> the decision on future nuclear new build, it is enormously important
> that the public have confidence in plans for dealing with nuclear
> waste and in the process by which these are reached.[37]

But Keith Baverstock, a radiation scientist from the University of
Kuopio in Finland – who was sacked as a member of CoRWM after a pro-
cedural wrangle, in April 2005 – was unimpressed by the Committee's
proposals. 'With a gestation period longer than that of an elephant, it
has delivered a mouse,' he said.[38] Time will tell if CoRWM or its critics
prove right.

Conclusion

Roger Helmer a Midlands Conservative MEP wrote in his local newspaper, 'As for [nuclear] waste disposal, I don't claim to be an expert, but the
British Geological Survey, in Keyworth, Notts, seems to think that deep
burial is the best bet, and I'm not arguing. This is the solution adopted
in Finland, and the Finns seem very happy with it.'[39] This is a common
observation, i.e. that Finland has cracked the nuclear waste problem.[40]
But has it? Looking at Finland may help us see what may lie in store for
the United Kingdom.

Certainly, both the controversially funded new Finish reactor – which
from the outset faced a challenge under EU State aids rules – and the parallel radioactive waste management programme installed at the same
site on Olkiluoto Island in south western Finland have been presented
by the Finish Government and European nuclear industry as the model
to follow.

Details of the history of the Finnish nuclear programme, its current expansion and waste disposal programme are available at several accessible web sites.[41]

But in summary, in 1992, Finland's Olkiluoto nuclear plant at Eurajoki began on-site shallow geologic disposal of low-level radioactive waste; and in 1998, low-level radioactive waste was disposed of at Finland's other nuclear plant site at Loviisa, east of the capital Helsinki.[42]

A division of TVO (Teollisuuden Voima Oy) – the company building the new adjoining reactor – called Posiva Oy is building the nuclear waste store, called Onkalo, some 500 m below ground using is a Swedish concept called KBS3, which Sweden's proposed repository also intends to implement.

As of mid-2006, the Finnish Government had yet to give final approval for the project located within the municipality of Eurajoki, but the entrance tunnel was nonetheless being constructed with the intention of it being the final store for all Finland's nuclear waste when it opens in 2020. Posiva Oy spokesman Veli-Matti Ammala asserted it would withstand the next Ice Age: something we will never know. And that is the rub with nuclear waste disposal – the plans have to cover geologic time, whereas humankind's longest lived institution is the Catholic Church, barely 2000 years old.

Compared to the almost universal 'Nimbyism' in the United Kingdom, in Finland local municipalities actually competed against other villages to be chosen as the site for the waste store because it would bring several hundred jobs and increase local tax revenues.

'When the site selection started in Finland, the nuclear industry said they would find the best geological site', commented Greenpeace Nordic energy campaigner Kaisa Kosonen. 'And, eventually, they chose the site on sociological reasons, because eventually Eurajoki was the first municipality to say "ok, we can take it", and there wasn't an active nuclear opposition in this area.'

The other key variable that has benefited progress with a final disposal strategy for nuclear waste in Finland is it only has four operational reactors, at just two sites, in a country with a comparatively low population density, mostly concentrated on the southern part of the country.

So, it seems unlikely that the experience there will be replicated in the United Kingdom. Instead, when it comes to specific sites being proposed, we are likely to see major confrontations, unless the industry can find a site where the local population is keen to have waste stored indefinitely.

You might think that the US proposal for a waste repository in a remote part of Nevada would face less problems . However the Yucca Mountain site has been the subject of a long-running battle.

The site has been identified as the favourite for the final disposal of very-long lived high activity waste for nearly two decades, but scientific disputes over the suitability of its geology and legal and political squabbles, led by an oppositionist State of Nevada, have resulted in perhaps the most protracted decision-making process on radioactive waste anywhere on the planet.

The simple message is that few people want to face the issue – and yet we have to find some place for the existing waste to go. The simple answer is that this might be somewhat easier if we were not proposing to produce more.

References and notes

1. Blowers, A., D. Lowry and B.D. Solomon (1991) *The International Politics of Nuclear Waste*, Macmillan press, Basingstoke and London.
2. Managing Radioactive Waste Safely (MRWS), Defra, 91 pages, September 2001 http://www.defra.gov.uk/environment/consult/radwaste/pdf/radwaste.pdf.
3. http://www.defra.gov.uk/environment/radioactivity/waste/index.htm
4. http://www.corwm.org.uk/content-248
5. Hansard, 8 November 2005: Column 298W.
6. Hansard, 22 March 2006: Column 392W.
7. 'Waste body has no say on new reactors', *Independent letters*, 30 January 2006; 'Debate urged on tackling future nuclear waste', *Financial Times*, 20 June 2006.
8. 'Paris "tipped off" over fears about London 2012 site', *The Times*, 15 July 2005.
9. 'Blair backs new reactors on existing sites', *Financial Times*, 18 May 2006.
10. www.defra.gov.uk/environ/fcd/futurecoast.htm
11. Hansard, 8 November 2005: Column 324W.
12. 'Nuclear sites identified', *Financial Times*, 11 June 2005.
13. 'Secret nuclear waste dumps back in debate', Tiscali news channel, 10 June 2005.
14. 'Nuclear waste agency selected dumps on the basis of political expediency', *Sunday Herald*, 12 June 2005; 'The nuclear option', *The Sunday Sun* (Newcastle), 12 June 2005; 'Nuclear dump site fear', *This is Scunthorpe*, 14 June 2005.
15. Hansard, 25 October 2005: Columns 295-6W.
16. Hansard, 15 November 2005: Column 1068W.
17. www.nda.gov.uk/documents/nda_final_strategy,_published_7_april_2006.pdf.
18. Hansard, 8 May 2006: Column 37W.
19. 'MoD ignores call to clean up radioactive coastal waste', *Sunday Herald*, 6 November 2005.
20. 'Plans for work to seal nuclear waste shaft', *Dundee Courier*, 13 October 2005.
21. 'Dounreay nuclear store is leaking', *Sunday Herald*, 19 March 2006.
22. 'We don't want a nuclear dump at our back door', *The Northern Times*, 6 January 2006.
23. 'McConnell: nuclear waste issue must come first', *The Herald*, 20 April 2006.
24. *The Guardian*, 9 January 2006.

25. 'Fear over future UK nuclear leaks,' *New Scientist*, 7 January 2006.
26. 'Cost of cleaning up after nuclear power stations are closed down rises to … £70bn', *Independent*, 3 January 2006.
27. The Long-Term Management of Radioactive Waste: The Work of the Committee on Radioactive Waste Management, Royal Society, January 2006.
28. 'Call for UK nuclear clean-up plan', BBC on line, 19 January 2006.
29. 'Danger: Nuclear Waste – Deal with disposal first, warn advisers,' *Independent*, 24 January 2006.
30. Lords Hansard, 16 February 2006: Column WA193.
31. Hansard, 16 February 2006: Column 1550.
32. 'US to clean up on UK nuclear mess', *Observer*, 5 March 2006.
33. 'Australia "should take nuclear waste"', News.com.au, 31 March 2006.
34. 'The man trying to lift the UK's nuclear cloud', *Sunday Times*, 2 April 2006.
35. CoRWM e-Bulletin, No. 5, April 2006, p. 1 www.corwm.org.uk/PDF/1697%20-%20e-bulletin%20-%2005%20-%20April%20-%202006.pdf.
36. CORWM e-bulletin No.7, May 2006 www.corwm.org.uk/pdf/1717%20-%20e-bulletin%207%20-%20May%2006.pdf. Explanation at: www.corwm.org.uk/pdf/1725%20-%20draft%20recommendations%20and%20rationales.pdf.
37. Conservative Party press release, 27 April 2006 www.conservatives.com/tile.do?def=news.press.release.page&obj_id=129461; Hansard, 4 May 2006: Column 1091.
38. 'Bury your nuclear waste, UK advised', *New Scientist*, 26 April 2000.
39. This is Leicestershire, letters, 30 May 2006.
40. For example, 'Finland buries its past', BBC on Line, 27 April 2006; 'How Finland fell back in love with nuclear power', *Sunday Telegraph*, 21 May 2006; 'Finns have waste problem wrapped up', *The Times*, 12 June 2006.
41. The Finnish nuclear waste management is guided by the Nuclear Energy Act and Decree (http://www.stuk.fi/ydinturvallisuus/ydinvoimalaitokset/saannosto/en_GB/saannosto/); Nuclear Energy in Finland, OCRWM, Washington DC, September 2005 www.ocrwm.doe.gov/factsheets/doeymp0410.shtml; World Nuclear Association briefing paper, 2005 www.world-nuclear.org/info/inf76.htm; Strength and Strain Anisotropy of Olkiluoto Mica greiss, Matti Hakala, Harri Kuula and John Hudson, October 2005.www.posiva.fi/tyoraportit/WR2005-61www.pdf. Further technical reports at: www.posiva.fi/englanti/tietopankki_tyoraportit.html?sstring=&avuosi=1990&lvuosi=2010&tyorap=on&sort=file&lang=eng.
42. Finland's Radioactive Waste Management Programme, STUK www.stuk.fi/ydinturvallisuus/ydinjatteet/loppusijoitus_suomessa/en_GB/loppusijoitus/.

10
Nuclear's Inherent Insecurities

David Lowry

> Public confidence about adequate nuclear security will
> be a pre-requisite for an expanded nuclear energy sector.
>
> Anita Nilsson, Head of Office of Nuclear Security, Inter-
> national Atomic Energy Agency, presentation to TopNux
> 2006 European Nuclear Energy Association conference,
> London, 22 March 2006.

One of the reasons ministers have given in support of new nuclear
plants is because the United Kingdom's fossil fuel supplies, especially
natural gas, in the future face several insecurities. Indeed, among the
lead reasons used to explain the need for the Energy Review held in the
first quarter of 2006 was fear of disruptions of security of supply of
energy sources.[1] The fact that this is not true is not as relevant as the
reality that it has become part of political lore.[2] Taking a contrary position,
this chapter demonstrates the insecurities inherent in nuclear power.

Nuclear insecurity – a national and international concern

> The safety and security of nuclear activities around the globe remain a
> key factor for the future of nuclear technology … The IAEA [also] has
> strengthened its co-operation on nuclear security issues with other
> international organizations, including the UN and its specialized agen-
> cies, Interpol, Europol, the Universal Postal Union and the European
> Commission.
>
> IAEA (International Atomic Energy Agency) Director General
> Dr Mohamed ElBaradei, 3 November 2003.[3]

The importance credited to nuclear safety by Dr ElBaredei in the above quotation is unsurprising. What is new and important is the coupling of safety with *security*. In his speech, he went on to emphasize:

> In light of the increasing threat of proliferation, both by States and by terrorists, one idea that may now be worth serious consideration is the advisability of limiting the processing of weapon-usable material (separated plutonium and high enriched uranium) in civilian nuclear programmes – as well as the *production* of new material through reprocessing and enrichment – by agreeing to restrict these operations exclusively to facilities under multinational control.

It is arguable that this is a remarkable statement coming from the director of the United Nations' agency charged with the promotion of the nuclear industry. But when it is considered that he made this comment barely two years after the terrorist attacks of 9/11 in the United States, it becomes more explicable. Dr ElBaredei – who was subsequently awarded the Nobel peace prize, the highest international honour, for his work done on nuclear security and nonproliferation – further pointed out that 'one convention that has gained increased attention recently is the 1979 Convention on the Physical Protection of Nuclear Material (CPPNM). In the past two years, 20 additional States have become party to the Convention, reflecting the importance of the international nuclear security regime.'

As a result of several terrorist events in Britain in 2005 – thankfully none involving radioactive materials – along with both government ministers producing anti-terrorist initiatives, and parliament debating their pros and cons such as the Prevention of Terrorism Act,[4] the recent past has seen terrorism rise to and remain at or near the top of the public and political agenda. This was reflected in similar attempts to tighten the international grip on terrorist threats. For example, UN Secretary-General Kofi Annan told an international conference of some 400 anti-terror experts – meeting in the Spanish capital on 10 March 2005 to mark the first anniversary of Madrid's own major terror attack on its suburban rail system – that terrorists must be denied the means to carry out a devastating nuclear attack. This is not a theoretical threat, but a very real one, as the revelations in the court case involving seven alleged terrorists in several plots between October 2003 and March 2004[5] – and other subsequent related revelations[6] – have made all too clear.

In urging UN member states to adopt the international convention on nuclear terrorism, Mr Annan made this blunt statement: 'Nuclear terrorism is still often treated as science fiction – I wish it were.' The secretary-general stressed: 'But unfortunately we live in a world of excess hazardous materials and abundant technological know-how, in which some terrorists clearly state their intention to inflict catastrophic casualties', adding: 'Were such an attack to occur, it would not only cause widespread death and destruction, but would stagger the world economy and thrust tens of millions of people into dire poverty.'

These high social stakes underline the importance of the need to ensure the best possible security applied to nuclear installations and materials.

Nuclear security worldwide

Shortly after the Madrid anti-terrorism conference, the world's nuclear watchdog, the IAEA held a timely international conference on 'Nuclear security: Global directions for the future' in London between 16–18 March 2005, co-hosted by the UK government. It was the culmination of over three years planning, as immediately after the terrorist events of '9/11' in the United States in September 2001, the IAEA General Conference – which always meets annually in Vienna in September – requested a review of the Agency's activities relevant to preventing nuclear terrorism.

The first day of the Nuclear Security Conference reviewed the achievements and shortcomings of international efforts to strengthen nuclear security; the second day explored how the international nuclear security regime is adapting to new measures – and the IAEA role in underpinning them; the third day focussed upon how the international community could reach a common understanding to better respond to the global threat of nuclear terror, detect and prevent it.

Richard Wright, the UK representative on the IAEA Board of Governors, summed the perceived situation after three days of intensive exploration by technical experts and decision-makers with words: 'Nuclear terrorism is one of the greatest threats to society.'[7]

The final statement of the conference made for chilling reading. It coolly declared: 'Priorities for strengthening nuclear security include continued efforts to enhance the prevention of terrorist acts; the physical protection and accountability of nuclear and other radioactive material in nuclear and non-nuclear use, in storage and transport, throughout the life cycle, in a comprehensive and coherent manner. A graded approach should continue to be used under which more stringent controls are applied for material or activities that pose the highest risk; for example, particular attention should be given to high enriched uranium or plutonium.' And added: 'The fundamental principles of nuclear security include embedding a nuclear security culture throughout the organizations involved. By the coherent implementation of a nuclear security culture, staff remain vigilant of the need to maintain a high level of security. The long-term sustainability of nuclear security efforts is a primary concern. The investments made in States, through their own efforts and through assistance programmes, must be sustained in order to continue to upgrade or maintain an adequate level of security. While the level of threat may change from time to time, an effective level of nuclear security must be appropriately maintained.'

The international nuclear watchdog is determined to keep vigilant, because its leadership and membership is all too aware of the negative implications for the nuclear generation industry were either a terrorist (or less likely, a wartime) attack to take place anywhere in the world against a nuclear power plant or nuclear installation, or illicitly obtained nuclear material to be used in a so-called dirty radiological bomb in a dense urban area.

Horrifying so-called dirty bombscenarios have been painted in several well-read weekly magazines. The sober finance sector magazine, *Business Week*, to mark the fourth anniversary of the '9/11' attack on New York, ran a frighteningly real account of what such an attack might mean for a densly populated urban area such as Manhattan.

At 8:30 a.m. on a Tuesday morning, as commuters converge on Manhattan, an al Qaeda operative explodes a dirty bomb outside the New York Stock Exchange. The device, while not especially powerful, contains a radioactive payload – in this case, cesium extracted from radiological equipment that was stolen from a New Jersey hospital by a sleeper working there as a lab tech.

The initial blast kills only a few dozen people, but radiation is quickly dispersed by the prevailing winds. Minutes after the explosion, New York City Police officers arrive – still unaware of the real nature of the blast. But when a radiation detector in one officer's car goes wild, it becomes clear that a dirty bomb has detonated in the financial center of America's biggest city ...[8]

It concluded....'Six months later, the financial district remains largely off-limits, and the local economy is limping along amid a cratering of business confidence, the collapse of the tourism industry, and a property market in free fall. Economists put the eventual economic losses at an astronomical $1 trillion....'

More recently, in March 2006, the internationally respected science and technology weekly, *New Scientist*,[9] carried a similarly frightening analysis of a similar nuclear terrorist scenario, under the chilling headline: 'Nuclear nightmare in Manhattan'

It pointed out that one serious problem would be the extent of the contamination from an uncontrolled dispersal of radioactive material 'for 150 kilometres or more downwind of the blast, dangerous amounts of fallout continue to drizzle down. …This nightmare scenario is one the US government is taking seriously. In the past two years alone, it has committed hundreds of millions of dollars to dealing with the aftermath of an act of urban nuclear terrorism, or a 9/11-style attack on a nuclear plant.'

The *Business Week* 'dirty bomb' scenario describes nuclear material stolen from a hospital, but there is real concern such material could be illicitly obtained from commercial nuclear power installations. UN Secretary-General Kofi Annan warned the Clinton Global Initiative, which ran contemporaneously (14–16 September 2005) with the UN World Summit in New York, of the dangers that proliferation posed, in giving terrorists opportunities to steal nuclear products that they could use to make so-called dirty bombs, which would combine radioactive material with conventional explosives in order to make bombs that could spread harmful radioactivity over a wide area. He reinforced these concerns in his address in the opening ceremony of the 60th General Assembly of the United Nations on 17 September 2005, when he insisted: 'we face growing risks of proliferation and catastrophic terrorism …'[10] So, for New York, read Paris, Tokyo, Moscow, London, Manchester, Glasgow, Birmingham or any other major urban centre.

An issue of outstanding consideration is that there is an apparent disconnect between the justifiable international concern over security threats posed by the insecurities of commercial nuclear energy sector – as the ongoing saga with Iran's claims and the international community's counterclaims over its nuclear programme amply demonstrates – and the promotion of an expanded nuclear sector by the very authorities who warn of the risk! It is indeed peculiar that, just as policy makers at the London and Madrid conferences were starting to face up to the real insecurities posed by existing nuclear plants and fuel cycle installations, and the transports between them, the IAEA should announce at the start of March 2005 that there were 'rising expectations for nuclear electricity production'. An official IAEA statement, issued in conjunction with publication of its *Nuclear Technology Review – Update 2005*, said: 'The IAEA forecasts stronger growth in countries relying on nuclear power, projecting at least 60 more plants will come online over the next 15 years to help meet global electricity demands.'

But the 'Faustian bargain'[11] – the now notorious nuclear deal famously spoken of by former Oak Ridge National Laboratory Director Dr Alvin Weinberg in 1972 – still exists. The spread of nuclear energy leads inexorably to the greater potential for the spread of nuclear weapons. The international spread of nuclear power technology, and its concomitant spread of nuclear materials, was no accident: it was heavily promoted under the US export technology transfer initiative for 'Atoms for Peace', set out in a path-breaking address to the UN General Assembly by President Dwight D. Eisenhower on 8 December 1953. In his address, in which he presaged the IAEA, Eisenhower commented:

> My recital of atomic danger and power is necessarily stated in United States terms, for these are the only in controvertible facts that I know. I need hardly point out to this Assembly, however, that this subject is global, not merely national in character. ...So my country's purpose is to help us move out of the dark chamber of horrors into the light, to find a way by which the minds of men, the hopes of men, the souls of men every where, can move forward toward peace and happiness and well being. ...Who can doubt, if the entire body of the world's scientists and engineers had adequate amounts of fissionable material with which to test and develop their ideas, that this capability would rapidly be transformed into universal, efficient, and economic usage.[12]

Nearly fifty years later we witnessed the near inevitable outcome of this policy of nuclear promotion: the Byzantine tale of deliberate – and covert – proliferation of nuclear technology by Abdul Qadeer 'AQ' Khan, former head of the Pakistan Atomic energy Commission, who carried out his secret proliferative deals while still holding this responsible post.[13] Yet when his successor addressed the IAEA nuclear security and terrorism conference held in London, mentioned earlier, he made no reference to Dr Khan's nefarious nuclear activities at all!

A few months before AQ Khan's nefarious nuclear network was exposed, the IAEA's Dr ElBaradei told a Carnegie Peace Foundation Conference on International Non-Proliferation held on 21 June 2004 in Washington, DC:

> Nuclear components designed in one country could be manufactured in another, shipped through a third (which may have appeared to be a legitimate user), assembled in a fourth, and designated for eventual turnkey use in a fifth. The fact that so many companies and individuals

could be involved is extremely worrying. And the fact that, in most cases, this could occur apparently without the knowledge of their own governments, clearly points to the inadequacy of national systems of oversight for sensitive equipment and technology.

On atomic exports he concluded that 'the present system of nuclear export controls is clearly deficient'.

He then argued: 'First, we must tighten *controls over the export of sensitive nuclear material and technology*. … Second, it is time that we revisit the availability and adequacy of controls provided over sensitive portions of the nuclear fuel cycle under the current non-proliferation regime… We should consider limitations on the production of new nuclear material through reprocessing and enrichment …considerable advantages – in safety, security and non-proliferation – would be gained from international cooperation in the front and the back end of the nuclear fuel cycle. Third, we should work to help countries stop using weapon-usable material (separated plutonium and high enriched uranium – HEU) in their civilian nuclear programmes…'. And he ended:'Fourth, we should eliminate the weapon-usable nuclear material now in existence.'

Warming to his theme as head of the world watchdog on atomic activities Dr ElBaredei wrote early in 2005 that nations should now seriously consider a five-year moratorium on building new facilities for uranium enrichment and plutonium separation: 'There is no compelling reason for building more of these proliferation-sensitive facilities, the nuclear industry already has more than enough capacity to fuel its power plants and research facilities', he wrote.[14]

One of the key events at the United Nations Global summit, held on 14–16 September 2005, was to open for signature the Convention for the Suppression of Acts of Nuclear Terrorism, as adopted by the UN General Assembly on 13 April 2005.The nuclear terrorism treaty – which strengthens the global legal framework to combat the scourge – requires the extradition or prosecution of those implicated and encourages the exchange of information and inter-state co-operation. It gained its first signatories in Russian President Vladimir Putin, US President George W. Bush, and French Prime Minister Dominique de Villepin. It enters into force thirty days after it is signed and ratified by at least 22 states.

In a related development at the United Nations, the United Kingdom submitted a resolution, Number 1624, on 14 September 2005 to the Security Council calling for tougher controls on terrorism, including nuclear threats. It stressed the Security Council 'calls upon all States to become party, as a matter of urgency, to the international counter-terrorism

Conventions and Protocols whether or not they are party to regional Conventions on the matter, and to give priority consideration to signing the International Convention for the Suppression of Nuclear Terrorism.' The Prime Minister's Strategy Unit had advised on this pressing concern earlier in a special study prepared for Tony Blair.[15]

US President Bush pressed for the Security Council to approve the resolution that called upon all nations to take steps to end the incitement of terrorist acts, and to commit countries to prosecute, and extradite, anyone seeking fissile materials or the technology for nuclear devices.

A powerful briefing issued by the UK Nuclear Free Local Authorities early in 2006 explored the likelihood of a future nuclear terrorist attack, based on past experience.[16] It listed several nuclear terrorism incidents, including one involving Britain's biggest and newest nuclear plant at Sizewell. Photographs, slides, maps and detailed information about types of radioactive materials and where they are stored were found in a car linked to one of the London terror suspects, in a raid after the July 2005 bombing campaign. The Metropolitan Police told one nuclear expert that sensitive material, which appeared to come from lectures and talks the expert had given in 2002, had been found in the car. More dramatically, a foiled Chechen rebel assault on the Russian city of Nalchik in October 2005 was reported to have involved an attempt to hijack five planes that could be flown into various targets, including a nuclear power station.

Indeed, so great is the dread risk of a terrorist attack on nuclear facilities perceived by some experts that one respected public policy institute, the Oxford Research Group, told the House of Commons Environmental Audit Committee Inquiry into 'Keeping the Lights on' that nuclear power should not be part of the UK's energy supply – precisely because it presents a major threat to our national and international security and increasing the risk of nuclear terrorism, by creating opportunities for terrorist organizations.

The National Fitness Leadership Alliance (NFLA) report helpfully collates details of several known previous physical threats to nuclear installations: To date, it is known there have been six direct attacks on nuclear power plants in France, South Africa, Switzerland, the Philippines and Spain (there may have been others which have not been made public). Fortunately, all of the reactors were in the early stages of construction and were not operational. The International Policy Institute for Counterterrorism (ICT) database includes some 167 terrorist incidents involving a nuclear target for the period 1970–99. Between 1966 and 1977 there were 10 terrorist incidents against European nuclear installations (reactors plus other types of nuclear facility). Between 1969 and 1975 there were 240 bombing threats against US nuclear facilities, and 14 actual and attempted bombings. According to a Russian intelligence official, during the years 1995–97 there were 50 instances of nuclear blackmail in Russia. Thankfully, most turned out to be hoaxes.

Nuclear insecurity in the United Kingdom

In September 1999, British Nuclear Fuels Limited (BNFL), then the key strategic nuclear company in the UK, as owner–operators of nuclear fuels cycle facilities such as Sellafield, and several ageing atomic power stations, launched an experimental attempt to create a dialogue with its so-called stakeholder – including its regulators, workforce, communities around its facilities, trades unions and some environmental pressure groups, including some from abroad. One of the outputs of this dialogue – now supercede by a national stakeholder dialogue run by BNFL's successor organization, the Nuclear Decommissioning authority – were two key reports on plutonium management options and security considerations, produced by two working groups of the BNFL stakeholder dialogue, each of which worked over several years, and involved high-level inputs from the BNFL Security Chief Dr Roger Howsley, and John Reynolds, deputy director of the government's security watchdog, the Office for Civil Nuclear Security (OCNS).[17]

The report of the Security Working Group – which produced 60 key recommendations – summarized its purpose as 'to contribute to the improvement of the security of BNFL's plant and activities, including in particular the transport of nuclear material, by the production of a quality review, using stakeholder dialogue, unique in this security context. The report is the fruit of rare collaborative effort on the part of a number of individuals from a variety of backgrounds with many differences in outlook. Notwithstanding that such differences in view were so divergent that in some instances they appeared to fully contradict each other, the group has produced what it considers to be a constructive and forward looking contribution to the manner in which security is provided for BNFL's activities.'

A number of differences on some security issues which were addressed in the course of the study remained unresolved, such as the manner of transportation of nuclear material, the risks arising from the release of sensitive information on nuclear materials, and just how much is it safe to make available to the public.

Contractorization and security controls

One issue looked at by the BNFL and more recent NDA stakeholder forums was that of the vetting of site operating personnel and drivers of vehicles transporting nuclear materials between licensed nuclear sites. It was argued by some participants that the competitive contractorization of the operational work of the NDA would undermine the ability of the OCNS to conduct proper vetting, as it will increase the number and dispersal of those contractors needing to be vetted. Here are some edited observations in respect of security vetting from director of the OCNS in his 2004 annual report:

> *Para 37. The vetting system works reasonably effectively, although it is unavoidably intrusive, time-consuming and labour-intensive. We advise foreign agencies that vetting is an essential component in nuclear security arrangements, in line with IAEA guidelines. There has been pressure in recent years to cut back numbers being vetted, but the current terrorist threat has brought about a prudent change of view.*
>
> *Para 38. It is sometimes claimed that a single government vetting agency could achieve greater efficiency. I doubt this. Our IOs are based close to nuclear sites. OCNS vetting staff are familiar with conditions within the industry, the hazardous nature of nuclear and radioactive material, and the work undertaken. If these close links were ever broken, the discernment of those undertaking interviews and record checks, and the understanding informing the decision-making process, would be lost.*
>
> *Para 39. The policy and practice of national security vetting gives full regard to the requirements of the Data Protection Act, the Freedom of Information Act, the Human Rights Act, race relations and other relevant legislation. Where appropriate, vetting information is exempt from disclosure. However, we are encountering some reluctance by employers and others interviewed to provide candid references. Despite assurances to the contrary, some worry that their identities might be disclosed on appeal and their organisations sued for defamation.*[18]

The subsequent fourth and fifth annual OCNS reports are available on the OCNS web site, part of the Department for Trade and Industry (DTI). The 2005 report informed the energy minister – to whom OCNS is responsible – that 'as part of a continuing programme of work since the terrorist attacks in the United States in September 2001, OCNS is reviewing, with the Nuclear Installations Inspectorate (NII) and the operators' own specialists, the security and safety measures in place to protect Vital

Areas at nuclear sites, including generating stations. Vital Areas contain equipment, systems or devices, the failure of which could have serious consequences for the secure and safe operation of a nuclear site.'

But on security vetting it revealed that 'the trends identified in previous years which show a steady annual increase in the numbers of individuals requesting clearance. Although no one has been permitted to work on a nuclear site without the appropriate clearance or escort, absorbing the increase in tasking has been a significant challenge; OCNS has had mixed success in meeting it. Against a background of staff shortages created by retirements on the one hand and a recruitment moratorium during a major DTI downsizing exercise on the other, OCNS productivity has continued to improve from an already high base as a result of better working practices and a focused commitment by staff on the task. In spite of this, a backlog of cases built up in the last two quarters of 2004 ....'

The OCNS director explained that his office provided a personnel vetting service which conformed to the requirements of the Nuclear Installations Security Regulations (NISR, 2003) and which applied to all permanent employees and contractors working in the civil nuclear industry. 'Clearances commensurate with the level of access to nuclear material and sensitive nuclear information are granted to individuals', he said.

The details are shown in Table 10.1, which includes totals recorded for 2002/2003 and 2003/2004 for comparison.

The figures confirmed at the end of the reporting period, the backlog (including arrears in revalidation cases) stood at 666 DV, 687 SC and 1600 BC+.

Details of the numbers of revalidation cases processed in 2004/05 are shown in Table 10.2 with totals for 2002/03 and 2003/04 included for comparison.

So, although overall the director general of OCNS said he was satisfied, he remained seriously concerned about backlogs, stating: 'The practical effect of the backlog on the industry amounts to delays in confirming

Table 10.1   Security vetting clearances granted by OCNS

Clearance levels	New cases		Revalidations	
	2002/03	2003/04	2004/05	2004/05
Developed vetting (DV)	312	435	471	279
Security checks (SC)	753	921	863	47
Counter terrorist checks (CTC)	22	23	1	0
Enhanced basic checks (BC+)	8,381	7,742	10,112	814
Totals	9,468	9,121	11,447	1,140

Table 10.2   Revalidation of security vetting clearances

Clearance levels	Revalidations		
	2002/03	2003/04	2004/05
Developed vetting	163	186	279
Security checks	51	101	47
Counter terrorist checks	0	0	0
Enhanced basic checks	2,552	3,059	814*
Totals	2,766	3,346	1,140

*Figure reflects policy change to treat BC+ (ie enhanced basic checks) revalidations as new cases.

recruitment of prospective employees or engagement of contractors, with the concomitant risk of losing them to other employment, and additional costs accruing from the need to escort individuals who do not hold the necessary clearance. In an industry where margins are so tight, this is a burden which must be reduced.'

Security vetting concerns were confirmed by the then Cabinet office minister, Douglas Alexander in a written answer: 'The most recent periodic review of the Government's personnel security system recommended the creation of a new official committee focussing on this area. That was accepted and that committee will work to ensure the continued effectiveness of personnel security policy and practice throughout Government, and in those organisations with which Government works in partnership.'[19]

It is now known from papers released in February 2005, under the Freedom of Information Act (FOIA), that in the year 2004–05 the OCNS had over 40 cases of potential security breaches under the current nuclear site management arrangements.[20] This should have provided food for thought for any ministerial evaluation of the implications of expansion of the nuclear energy programme in the United Kingdom, a point made by Labour's environment campaign ginger group, the Socialist Environment & Resources Association (SERA) in its evidence to the Energy Review in April 2006.[21]

The papers released under a FOIA request included the following information:

- OCNS carried out 129 pre-notified inspections in 2004. (These include the pre-notified inspections covering the last quarter of the FY 2003/4 cited in the last annual report.)
- OCNS has carried out 15 no-notice inspections since September 2003. *That OCNS does not know how many security passes were lost/stolen within the nuclear industry in the past year*. OCNS states it is primarily interested

in confirming that the sites have an effective pass management system as part of their security arrangements. This should include procedures for recording lost/stolen passes, disabling their access where access is automated, and periodic redesign and re-issue.

- OCNS does not expect to receive reports of individual alarms etc. where the cause has been investigated and assessed as benign. Nor are companies required to report what are essentially maintenance matters provided these have been dealt with promptly and have not significantly downgraded overall security effectiveness. We have interpreted your question broadly to mean anything out of the ordinary that the Operators drew to our attention. Therefore, not all of the incidents reported to us and noted here are of equal seriousness. None of the incidents involved theft of nuclear material or sabotage.

Every citizen – let alone minister – should remain concerned over the manifest inadequacies of the security arrangements covering the civil nuclear sector. And there is yet one more outstanding security concern.

Terrorist threat – could it determine the nuclear waste management option chosen?

The outputs of one of the CoRWM-sponsored expert workshops (see chapter 9) aimed at 'scoring' comparative hazards of different options in handling nuclear waste, that had worked up the ire of the Royal Society, were quietly released in a 187-page report posted on the CoRWM website without fanfare or media recognition on 11 January 2006.[22]

The security experts' workshop recommendations warned that Britain's nuclear waste was – and indeed is – vulnerable to terrorist attack and the government was failing to address the issue with sufficient urgency. It was not reported more widely for three months, in the *Daily Telegraph*[23] accompanying a CoRWM three-day public workshop that consolidated its draft recommendations to ministers. The nuclear security specialists – including the author of this chapter – convened by CoRWM to advise on the security dimension of the proposals for radioactive waste management over the next 300 years, included in their conclusions and recommendations[24]:

> It is our unanimous opinion that greater attention should be given to the current management of radioactive waste held in the UK, in the context of its vulnerability to potential terrorist attack. We are not aware of any UK Government programme that is addressing this issue

with adequate detail or priority, and consider it unacceptable for some vulnerable waste forms, such as spent fuel, to remain in their current condition and mode of storage.

The experts added: 'We urge the Government to take the required action and to instruct the Nuclear Decommissioning Authority, in cooperation with the regulators, to produce an implementation plan for categorising and reducing the vulnerability of the UK's inventory of radioactive waste to potential acts of terrorism, through conditioning and placement in storage options with an engineered capability specifically designed to resist a major terrorist attack.'

CoRWM's draft recommendations (April 2006)[25] made specific reference to the security concerns raised, stating: CoRWM recommends a staged process of implementation, incorporating the following element:

a. A commitment to the safe and secure management of wastes through the development of an interim storage programme that is robust against the risk of delay or failure in the repository programme.

Due regard should be paid to:

• reviewing and ensuring security, particularly against terrorist attacks;
• minimising the need for re-packaging of the wastes; and
• addressing other storage issues identified during CoRWM's public and stakeholder engagement process, such as avoiding unnecessary transport of wastes.

The security experts' conclusions also highlighted the danger posed by liquid high activity waste from the reprocessing of nuclear fuel currently stored at Sellafield in Cumbria. A spokesman for the OCNS said it was convinced that the procedures for protecting civil nuclear installations and processes were 'robust and fit for the purpose'. Interestingly, OCNS's deputy director was a member of the experts' workshop, and a signatory to the unanimous recommendations.

Other more widespread concerns over the insecurities on nuclear waste transported by train have emerged again since the start of 2006 – although there have been periodic regional concerns expressed for over the preceding 25 years. In early April 2006, Greenpeace UK released a study, prepared by experienced nuclear issues consultant John Large, on the potential hazards posed by the transport of spent fuel in the UK.[26] Consultant nuclear engineer Dr Large argued that a nuclear waste transport incident, such as a terrorist attack, could spread radioactivity over 100 km and cause over 8000 deaths, according to an internationally renowned nuclear engineer. Greenpeace argued that as the train routes pass through several large towns and cities, such as London, Bristol and Edinburgh, tens of thousands of people could be exposed to radiation in such an incident. The review concluded alarmingly that the transportation flasks containing spent nuclear fuel 'provide no extraordinary safeguard against terrorist attack' and would be at their weakest if caught in 'the high and sustained temperatures involved in a tunnel fire'.

In the United States, anti-nuclear activists characterize the hazards posed by the rail movement of nuclear materials presenting the public across the continent with a threat of a 'Mobile Chernobyl'. That equally applies in the UK. As the nuclear decommissioning programme gets underway in earnest over the next decade, necessitating a significant increase in nuclear transports, the issue of the safety and security of such transports will inevitably once more become one of widespread public concern.

Conclusion

There are security issues associated with each phase of the nuclear fuel cycle. Enrichment and reprocessing operations attract particular concern since they can involve the production of materials which can be used in weapons. But equally, waste storage and the transport of nuclear materials present possibilities for direct attack or theft, while nuclear facilities represent potentially attractive high-profile targets for terrorist assault. In a world where terrorism is on the increase, it would seem foolish to offer more targets, particularly by increasing unnecessary transports in plant decommissioning, and indeed more tools.

Certainly, given that attempting to minimize these problems will, at the very least, increase costs and disrupt fuel cycle and power plant operations, then, as the present author concluded in extensive evidence on behalf of the OU Energy and Environment Research Unit to the Commons Environmental Audit Committee's 'Keeping the Lights On' Inquiry, overall, far from enhancing energy security of supply, the further deployment of nuclear technology would undermine security in the UK, and should be avoided.

References and notes

1. Our Energy Challenge – Securing clean, affordable energy for the long-term – Energy Review consultation document: www.dti.gov.uk/files/file25079.pdf.

2. For example, the Energy Minister Malcolm Wicks wrote in his foreword to the Energy Review 'The UK has become a net importer of gas sooner than expected.' Yet, the same minister told parliament in a written answer in February 2006 that while the 2003 Energy White Paper did not specifically forecast the rate of depletion of UK gas, it did say that 'it is ... likely that the UK will become a net importer of gas on an annual basis by around 2006' (paragraph 6.13). He explained that this was 'in line with the projections of outside analysts'. For example, he added Wood Mackenzie, in its August 2001 multi-client report entitled 'Running sort of gas: The outlook for UK and Irish gas markets', had said 'It is probable that the UK will become a net importer of gas in either 2005 or 2006.' (Hansard, 7 February 2006: Column 1073-4W).
3. Statement to the 58th Regular Session of the United Nations General Assembly by IAEA Director General Dr Mohamed ElBaradei, 3 November 2003.
4. For example, Prevention of Terrorism Act 2005; Statutory Instrument (S.I.) No. 350 Prevention and Suppression of Terrorism, The Terrorism Act 2000 (Continuance of Part VII) Order 2005, 15 February 2005; S.I. No. 1525 The Terrorism (United Nations Measures) Order 2001 (Amendment) Regulations 2005; The Terrorism Act 2000 (Proscribed Organisations) (Amendment) Order 2005, 13 October 2005.
5. 'Trio charged with terror crimes', BBC on line, 1 October 2004.
6. For example, terror suspect linked to 'nuclear bomb plot', *Guardian*, 22 March 2006; 'Testers slip radioactive materials over borders', *New York Times*, 28 March 2006.
7. The full findings of the conference are available on the IAEA web site at www.iaea.org/NewsCenter/News/PDF/conffindings0305.pdf.
8. 'New York takes another hit – It has been readying itself for a dirty bomb since 9/11, but ...' *Business Week*, 19 September 2005 www.businessweek.com/magazine/content/05_38/b3951012.htm.
9. 'Nuclear nightmare in Manhattan', *New Scientist*, 18 March 2006.
10. www.un.org/webcast/ga/60/statements/sg050917eng.pdf
11. 'Social institutions and nuclear energy', *Science*, 7 January 1972, pp. 27–34.
12. Part of the Milestone Document Series published by the National Archives & Records Administration, www.eisenhower.utexas.edu/atoms.htm.
13. 'The merchant of menace', *Time Magazine*, 14 February 2005, www.time.com/time/covers/1101050214/.
14. 'Curbing the nuclear threat', *Financial Times*, 2 February 2005, www.iaea.org/NewsCenter/News/2005/npt_2005.html.
15. Investing in prevention – an international strategy to manage risks of instability and improve crisis response (see section on the threats from WMDs) www.strategy.gov.uk/downloads/work_areas/countries_at_risk/report/index.htm.
16. www.nuclearpolicy.info/docs/scotland/Briefing_NFLA(S)_Jan06.pdf
17. Plutonium Working Group (196 pages, 31 March 2003) www.envcouncil.org.uk/docs/SWG%20Final%20Report.pdf; Security Working Group (145 pages, 10 December 2004) www.envcouncil.org.uk/docs/SWG%20Final%20Report.pdf.
18. OCNS third annual report at www.dti.gov.uk/energy/nuclear/safety/dcns_report3.pdf.
19. Hansard, 21 July 2004: Column 333W.

20. www.dti.gov.uk/about/foi/documents/ocns.pdf
21. www.sera.org.uk/publications/energyreview_members.htm
22. www.corwm.org.uk/PDF/1502%20-%20Overall%20Specialist%20 scoring%20report%20V1.1.pdf
23. 'Britain's nuclear waste vulnerable to terrorist attack', *Daily Telegraph*, 27 April 2006.
24. ibid. 22. CoRWM Specialist Workshops – Scoring, 6th,7th,8th,13th & 14th December 2005, Report: COR004.
25. www.corwm.org.uk/pdf/None%20-%20CoRWMs%20Draft%20 Recommendations%2027%20April.pdf
26. Risks and hazards arising in the transportation of irradiated fuel and nuclear fuel materials in the United Kingdom, 4 April 2006 www.greenpeace.org.uk/ climate/climate.cfm?CFID=4510607&CFTOKEN=79410701&UCIDParam= 20060404140317.

11
Risk, Economics and Nuclear Power

Catherine Mitchell and Bridget Woodman

Introduction

This chapter analyses the key areas that will have to be addressed to make nuclear power an attractive investment. It assesses the cost of nuclear power – something always difficult to pin down. It explains the various risks involved in building a new nuclear power plant – whether technological, investment, market, institutional, political, financial, capacity-related and so on. It also looks at what has been done in Finland and the United States to support nuclear power. Overall, it argues that the economics of nuclear power are so uncertain and the risks of delivering meaningful levels of new nuclear power plants so great, that it is far less risky and far more rational to develop other non-nuclear energy options.

There are a range of different levels of support required, depending on whether the goal is one or two new nuclear stations or a full nuclear power programme. Different actors – in particular government and private companies – have different goals and therefore different requirements. An overarching framework to reduce risk in line with a goal of delivering a new nuclear power programme would, in principle, be attractive to the industry. However, this is unlikely to be popular with all the UK's incumbent energy companies, which have worked hard to carve out their market share. Moreover, it is unlikely to be popular with parts of government as well as with all customers and consumer groups, which will wish to ensure that the best long-term outcome for customers is achieved.

The economics of nuclear power

The generating cost of nuclear power is a central factor in determining whether or not nuclear power is supported. Certainly the issue has been emphasised by those advocating new nuclear stations, particularly in comparison with the costs of supporting renewable technologies.[1] However, in an absolute sense, generating cost is relatively unimportant once a decision has been taken to create a nuclear support programme, since the effort in getting stations built will essentially centre on supporting it in such a way that it is built, despite its costs. The cost of doing this for the UK taxpayer will be effectively open-ended, because limiting amounts or period of support would inject unacceptable risk for investment.

The cost of the guarantees by government will largely be defined by the cost of constructing the reactor and managing its nuclear waste during operation and after closure. As can be seen from Tables 11.1 and 11.2, there is no consensus about these costs, although past experience shows that nuclear companies substantially underestimate costs.[2] The tables shows the degree of uncertainty about the costs of generating power from new design reactors, and also the extent to which they are dependent on key factors such as the project's operating lifetime and its load factor.[3] Load factor in particular is an important issue influencing the cost of electricity from the reactors and there is little evidence for the optimism of some studies that 90% load factor will be achieved from untested reactor designs.[4] Figure11.1 shows how the generating cost is related to the discount rate used in the various studies.

Investment risk

The Energy Minister Malcolm Wicks has claimed, categorically, that any future nuclear industry would have to operate without heavy government subsidy.[8] Despite this optimism, it is more generally accepted that in order to attract investment to build new nuclear power stations, the government would have to provide both direct and indirect subsidies.[9] This is supported by a number of recent reviews of nuclear power as an investment option.[10] These subsidies would have to perform a range of functions:

- Provide appropriate rates of return expected by investors
- Remove the risk of future political changes
- Minimise the costs of financing construction debt
- Ensure a market (i.e. a buyer) for the electricity when the station is finally commissioned
- Guarantee in particular that the nuclear waste and decommissioning liabilities of the plant's operation were minimised and/or capped.[11]

If the government is to be confident that it can attract investment for one nuclear reactor, let alone a programme, it will have to reduce investment

Table 11.1 Comparison of assumptions in recent forecasts for new nuclear power plants[5]

Forecast	Construction cost (£/kW)	Construction time (months)	Cost of capital (% real)	Load factor (%)	Non-fuel O&M (p/kWh)	Fuel cost (p/kWh)	Operating life (years)	Decommissioning schedule	Generating cost (p/kWh)
Sizewell B[6]	3,500	86		84	1.15	0.7	40	Part segregated, part cash flow	
Rice University	~1,300								5.0
Lappeenranta University			5	91	0.5	0.2	60		1.6
Performance and Innovation Unit	<840		8 / 15	>80			15 / 15		2.31 / 2.83 / 3.79
Massachusetts Institute of Technology	1,111	60	11.5	85	0.83[7]		40		3.7
Royal Academy of Engineers	1,150	60	7.5	75 / 90	0.45	0.4	25 / 40	Included in construction	4.4 / 2.3
Chicago University	555 / 833 / 1,000	84	12.5	85	0.56	0.3	40	£195 million	2.9 / 3.4 / 3.9
Canadian Nuclear Association	1,067	72	10	90	0.49	0.25	30	Fund (0.3 p/kWh)	3.3
IEA/NEA	1,100–2,500	60–120	5	85	0.38–0.90	0.15–0.65	40	Included in construction cost	1.2–2.7
OXERA	1,625 first plant		10	95	0.35	0.3	40	£500 million in fund after 40 years life	1.8–3.8
	1,150 later unit								

Table 11.2   Generating cost estimates for new nuclear plants

	MIT 2003	PIU 2002	Chicago University 2004	RAE 2004	DGEMP 2005	Finland 2003	OECD 2005	OECD 2005
Generating cost (p/kWh)	3.9–4.0	3.0–4.0	3.1–3.6	2.26–2.44	2.0	1.7	1.3–1.9	1.8–3.0
Rates of return (%)	11.5	8 and 15	12.5	7.5	8	5	5	10
Capital cost ($/kW)	2000	2000	1500	2000	1413	1900	1000–2000	1000–2000
Load factor (%)	85	75–80	85	90	90	90	85	85
Economic life (yr)	15	20	15	25 and 40	35–50	40	40	40
Construction period	5	n/a	5–7	5	5	5	4–6	4–6

Notes: MIT: Massachusetts Institute of Technology; PIU: Performance and Innovation Unit; RAE: Royal Academy of Engineers; DGEMP: General directorate for energy and raw materials; OECD: Organisation for Economic Co-operation and Development; n/a: not applicable.
Source: From BNFL Memorandum to the EAC, page 161.

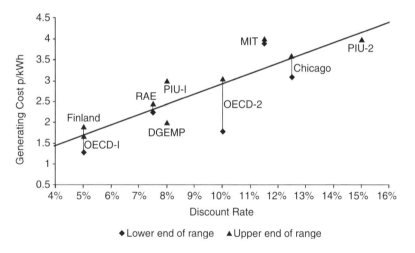

Figure 11.1   Generating costs vs discount rates (BNFL Memorandum to the EAC, page 162)

risk to the point that investors feel confident that they can be sure of a return whatever the changes in political parties and policies.[12]

Any framework support mechanisms for new nuclear power will have to be acceptable under European State Aid rules. This chapter has taken the view that government will be able to meet State Aid requirements. If it cannot, then it will be very difficult for a nuclear power programme to be established. Establishing what it can, or cannot do, in support of nuclear power therefore has to be a matter of urgency for government. Without such assurances, the construction of nuclear power plants has to be left to the commercial wisdom of companies, none of whom have so far brought forward plans to build plants, despite the theoretical ability to do so.

Government measures to reduce investment risk come at a cost, and the more risk is reduced the more costly it will be. In financial terms risk would have to be reduced or removed in most if not all areas which could materially affect the return on investment. The extent and impact of these measures is unknown. Types of investment risk are set out below:

- Possible delays or risks in gaining planning permission
- Unexpected difficulties in construction leading to delays in commissioning
- Escalations in the cost of waste management and decommissioning
- Technology risks – for example, lower output than expected – and any technology licensing or other permitting concerns
- Market risks
- Revenue risks
- The need to cap a nuclear operator's liabilities in the event of a catastrophic accident
- Company risks (sale of company/takeover; strategy change; ability to sell nuclear stake)
- Capacity risk – the ability to deliver a programme of nuclear power plants, including being sure of enough skilled employees
- Political risk
- Policy risk
- International political risk.

Given the scope of these risks, reducing investment is not the simple task of simply paying a large (enough) subsidy – it requires a complex, inter-linked basket of support. Investors in any new project have to be reasonably certain that they will realise an adequate rate of return on their outlay. The introduction of liberalisation and competition increases the risk that returns will not be sufficient to attract investment. Returns on investments in electricity utilities have been in the range of 7–12% since privatisation.[13] The required rate of return for a new nuclear project is uncertain: one view is that because of the complex and long-term nature of the risks associated with nuclear projects, the rate of return required to stimulate investment is more likely to be around 14–16%.[14] Another view is that reducing risk would bring rates of return down towards the average figure.

In order to enable new build, the industry has to convince investors that the risks of a new nuclear project are sufficiently reduced or mitigated and that the investment will earn an acceptable rate of return. The nuclear industry's submissions to the 2006 Energy Review set out some of the preferred measures[15]:

- Prelicensing of reactor designs
- Limiting public inquiries into a project
- Ensuring that the market is designed to value the low carbon operation of nuclear stations, and guaranteeing this situation for the long term. This could be in addition to the existing EU Emissions Trading Scheme, which, as a market mechanism, does not provide the necessary long-term certainty about carbon prices to encourage investors in new nuclear projects. The preferred mechanism seems to be a carbon contract offering government payments for carbon reductions
- Seeking a commitment that the government will take over the financial and management responsibilities for nuclear wastes, including spent fuel, after a specified period of time.

These options are geared towards providing certainty for investors as well as giving direct, or indirect, operating subsidies to the industry. Even if only a few of these demands were set in place, it is clear that significant changes to the structure of the competitive electricity market would be needed to allow financing of new nuclear stations to maintain nuclear generation at around 20% of the UK's requirements. This would exacerbate the indirect market distortions created by the open-ended subsidies already granted to the nuclear industry as a result of the bailout of British Energy in 2003.[16]

Whatever the combination of support mechanisms chosen, government would effectively have to subsidise all stages of a nuclear reactor's life – construction, operation and the open-ended decommissioning and nuclear waste management phase after the plant was closed.

Market risk

Liberalised electricity markets have proven to be a difficult environment for nuclear power plants. This is because of nuclear station's long construction times, high construction costs, inflexibility in generation and uncertainty about the costs of dealing with nuclear waste – all of which add risk to the project. When compared to natural gas power plants which can be of almost any size, can be constructed quickly, and are cheaper and flexible generators, it is little wonder that a new nuclear power plant has not yet been built and operated in a liberalised electricity system.[17]

Electricity systems operate in real time. In other words, at any given moment, the supply of electricity has to equal demand. The level of demand fluctuates according to the time of year and the time of day, with the peak occurring on a cold winter's weekday at around 5 p.m. There is, however, a minimum amount of electricity which always has to be supplied, known as the base load.

Nuclear power stations are inflexible generators as they are unable to follow the peaks and troughs of demand, and instead have to operate at a constant level of output. The reasons for this are partly technical: increasing and decreasing temperatures to adjust output causes thermal stresses which over time can lead to components cracking or even breaking. Repairing, or replacing degraded components would require the plant to be closed. Nuclear stations therefore supply base load power, leaving more flexible generators to 'load follow' as demand increases or decreases. By providing base load electricity, the generator is able to operate constantly at maximum output, thereby minimising the cost per kilowatt.

Given the high costs of nuclear projects, investors have to be confident that any nuclear station will be able to operate in the most cost-effective way to provide certainty that they can finance their debts from the project's construction. Investors will therefore need to be confident that the government will ensure that market and revenue risk is removed so that the station's output can be sold at an acceptable price. This means that government must be prepared to underwrite, or guarantee, prices for new nuclear output through the market, either directly or indirectly, for several decades.

There is no guarantee that the capital and decommissioning costs of new build nuclear power plants will ensure that they are economic at base load, even if their generation costs are low. In order to ensure that they will be able to operate at base load, nuclear plants would need a range of support mechanisms, as outlined earlier, to:

- bring the price per kWh down, either by subsidies or framework conditions, to a competitive level so that the plants can operate within the electricity market at base load;
- ensure that bilateral contracts between the generator and suppliers guarantee buying all electricity produced, allowing plants to run all the time, even if this is more expensive than other base load electricity; or
- directly alter the market to enable nuclear power to operate as a base load generator.

The fact that such a large proportion of electricity would be supported in an open-ended way would mean that the electricity market would reduce substantially in size. Competition in generation will therefore be reduced to a far smaller market.

The nuclear industry emphasises its role as a base load generator, but tends not to acknowledge the inevitability of this role given the technology's relative inflexibility. The point here is that by providing a high proportion of the UK's base load demand, the nuclear industry is denying other generators the opportunity to do so. This in turn will leave other generators with uncertain markets for their output, meaning that investors will be less confident of receiving a return on their investment in these other projects, and are therefore less likely to put their money in.

New reactor designs are theoretically more flexible than existing reactors currently operating in the UK system. Even if operators did decide to increase and reduce output in response to demand, this would have an adverse effect on the economics of the station. As pointed out earlier, load factor is an important element in the costs of power from reactors, and in order to achieve as low a cost as possible, operators will need to be operating the highest load factor possible for as much of the time as possible. While operating more flexibility may be technically feasible for new reactors, it will not be economically attractive.

In addition, the electricity market is an 'energy' or electricity 'only' market meaning that it is solely market prices which are meant to prompt the correct workings of the market, including providing incentives for new capacity. The ability of market prices to prompt appropriate capacity responses will be dulled if 30–40% of the market is ring-fenced through obligations covering both renewable and nuclear output, and for baseload generation. This is qualitatively different from ring-fencing renewables generation because renewables are projected to fall in price, become competitive and move into the market. Nuclear power would have to be ring-fenced for the lifetime of the power plants, which is at least 40 years.

The UK government is arguing for liberalisation, competition and transparency of prices in areas such as the international and European natural gas market. The success of privatisation and liberalisation has been the catalyst to making costs and prices more transparent, although those related to nuclear power still remain shrouded in uncertainty. That the government is considering moving backwards within the electricity market with such potential long-term consequences is an extremely retrograde step.

Government risk

There are considerable risks to government for supporting a nuclear power programme. Therefore investors will perceive there to be considerable political risk. A nuclear power plant will live through at least ten or a dozen parliamentary lifetimes, from discussion to decommissioning. There is a real risk that both policy and the political situation will change. Investors will require confidence that their investment can be recovered irrespective of this.

Investors will take the history of energy policy and nuclear power development in the United Kingdom into account when making decisions. In 1979, the then Conservative government announced that there would be a programme to construct ten new reactors, with the construction of one reactor beginning each year from 1982.[18] In the event, only one was built, Sizewell B, which came on line in 1995. This was in a monopoly situation and should have been a far easier task than it would be to build nuclear reactors today with more technologies, more customer involvement and liberalisation. Subsequent White Papers in 1995[19] and 2003 found against nuclear power on economic grounds. Within 3 years, another review is considering the question again. Given the controversy behind this review, investors must ask: 'how long will it be before energy policy is reviewed again?'

The risks to government fall into the following categories:

- Supporting a technology that may be found to be more expensive than renewables and demand reduction.

  The costs of different forms of electricity generation in 2020 are unknown. Projections behind the Performance and Innovative Unit's (PIU) Energy Review and the 2003 Energy White Paper (EWP) showed that some renewable technologies were projected to be among the cheapest generating technologies by 2020, as their prices fell compared to gas. Some technologies (onshore and offshore wind, some biomass and possibly some wave) were projected to be cheaper than nuclear power.[20]

- Getting 'locked-in' to a technology which may be expensive over the long-term; which does not contribute to government goals of sustainability, security and fuel poverty reduction; and whose success will take a long time to evaluate.

  Support for nuclear power will almost certainly have to cover all aspects of the project's life. While day-to-day generating costs may become competitive (e.g. because of increasing gas prices), nuclear would still require a surrounding package of support. In the event that individual renewable technologies do not become increasingly competitive, support for them can be curtailed relatively easily given the high number of small projects and the mixture of technologies. Support for a nuclear power programme is 'locked in' to a much greater degree, not only because of the time taken to construct and ascertain whether it is working or not, but also because of the need to provide support for dealing with the nuclear waste it will have created.

- By supporting a mature technology, that is not projected to fall in price, the government is effectively jettisoning principles of public policy expenditure.

    The extent of support outlined above for new nuclear build is qualitatively different from support needed for a renewables programme, since support for renewables technologies is projected to decline as those technologies become competitive, in line with the principles of public policy investment. Public expenditure on technologies is generally aimed at stimulating innovation, developing options and is under the expectation of price falls. As far as possible, public money should not interfere with the direct functioning of the market unless part of a clear innovation policy.

- Announcing a review of energy policy so soon after the last Energy White Paper has focussed attention on the issue of political risk, thereby potentially undermining or delaying investment decisions while energy policy is effectively in limbo.

There is a very real risk that actions by the government to reshape the market will undermine investment, not just in renewables and demand reduction but also in natural gas heat and power plants. If the new nuclear programme does not perform, the UK's emissions reduction programme would be in serious trouble.

Energy security risks

The underlying energy policy argument of the 2002 PIU Energy Review and the 2003 Energy White Paper was that natural gas would act as a transition fuel to a low carbon energy system. Its percentage of the market would gradually decrease until it assumes a role as a flexible, 'balancer' on the system to complement intermittent electricity from renewables.

However, increasingly, concerns about the security of natural gas supplies[21] have been used as an argument for new nuclear plant build. For example, the Confederation of British Industry (CBI), has voiced concerns that we are over-dependent on natural gas for electricity generation, domestic heating and industrial use and that renewables and demand reduction measures are not delivering fast enough to ensure that there is sufficient future capacity. The CBI is keen to maintain energy security – meaning enough electricity and gas for the needs of its membership at an acceptable cost. As part of this, it argues for rapid clarification of the place of nuclear power in the energy mix and progression of supportive measures for nuclear.[22]

Concerns about the future security of gas supplies are not as clear-cut as it might seem from the CBI's statements. The increasing availability of gas storage infrastructure – both storage and interconnectors – should do much to insulate the UK against the price spikes seen in the winter of 2005–2006, as well as providing a hedge against short-term supply problems. In addition, the United Kingdom is able to source its gas supplies from a number of different countries, both inside and outside Europe, so providing a degree of additional security through diversity. While it is unlikely that in the long-term gas prices will fall back to the very low prices seen in the 1990s, it is also very unlikely that gas supply will be interrupted or that there will be insufficient future capacity.

Institutional risk

Institutional resources will be needed to ensure that framework mechanisms are put in place to deliver a programme of nuclear power plants. This might require changes to legislation; the merging or de-merging of different departmental responsibilities and certainly a large increase in civil servants, particularly if public opposition becomes pronounced. While preparing for a nuclear power programme, it will be necessary to focus even more strongly on the promotion of renewables and demand reduction to ensure that no undermining occurs.

Institutional resources will be needed in the following areas at least:

- Planning: Action to reduce the risk of long, costly nuclear plant planning applications and the speeding up, or changing, of licensing procedures.
- Regulation: Ofgem, the energy regulator, is responsible for regulating the energy system. Ofgem works to Duties. These Duties would have to be changed if support for nuclear power were perceived to counter the current rules. More likely, providing there was no direct intervention in the market, Ofgem would be required to support the indirect developments and mechanisms of support via legislation. If there were direct

intervention, the fundamental attitudes to markets, as discussed above, would have to change.

- Departmental resources: All the potential direct and indirect support mechanisms discussed previously, such as planning, insurance liabilities, waste and decommissioning agreements, development of obligations, new legislation and liaising with the EU on State Aid would have to be undertaken through various institutions and departments. All would require additional resources. These institutions would include the Office of the Deputy Prime Minister for planning questions, the Environment Agency, the Nuclear Installations Inspectorate and Ofgem. Thus, it would require extensive civil service resources, a considerable challenge, particularly following the recent drive to reduce the size of the civil service.[23]

These are the institutional requirements of delivering a new nuclear programme. There are risks attached to all of them:

- There could be significant opposition to this if the fundamental planning procedures of the United Kingdom are threatened.
- The market and financial support required for a technology not clearly cheaper or better than alternatives could also be extremely unpopular and will also risk unravelling the very carefully built-up arguments for government expenditure and innovation.
- There is a risk that the government will not be able to bring the different departments together to deliver such a programme, particularly given it was unable to do so in the past, and when it has been so unsuccessful, for example, in the task of trying to link supply and demand by bringing the DTI and supply side together with Defra and the demand side.
- Public support has to be in place for the development of any new nuclear power stations. It has been such a long time since new nuclear stations have been built that a large proportion of the population has never been involved in any public debate about its desirability or otherwise. Public support for nuclear power is very uncertain – a recent MORI poll for the Tyndall Centre found that a majority of the public would support the development of renewables and energy efficiency over new nuclear stations as an option for addressing the UK's carbon dioxide emissions.[24]

Putting in place a set of conditions to encourage investment in nuclear power would be a substantial commitment. It would require an unknown, open-ended, commitment to support the technology to ensure that investors were insulated from the disadvantages of investing in this field. This would undo many of the moves to make electricity costs more transparent and could include ring fencing a portion of the electricity market to protect nuclear, an already developed technology. The experience of other countries, as discussed later, supports the argument that simulating a nuclear construction programme will require significant resources.

The knock-on effects of developing such a market would be significant for investors in other technologies, because much of the operating risk in the market would be shifted to new, non-nuclear projects. The situation could be particularly serious for developing technologies such as renewables, which rely on their lack of carbon emissions to attract investment. With the prospect of large quantities of nuclear generation scheduled to come on line in the future, the incentive to invest in new renewable technologies would be removed or, at best, watered down.

Risk of undermining other non-nuclear technologies

There would be considerable risks to the government in pursuing a new nuclear programme, largely because it would undermine investment in other technologies or low carbon options. The arguments in support of a new nuclear programme are based on the implicit assumption that all low carbon technologies and options are complementary: in other words that there can be a thriving nuclear power sector and continued emphasis on centralised generation operating in harmony with energy efficiency, renewables and other low carbon technologies.[25] If this is not the case, as this chapter argues below, and development of new nuclear plants fails, then the United Kingdom will be in a far worse situation than it currently is.

The resources brought together to support nuclear power will provide policy mechanisms to reduce risk and provide investor confidence. The mirror effect of this is that as the resources support nuclear they undermine the development of the low carbon energy system, through a combination of:

- Undermining the institutional resources required for developing renewables and demand reduction. Existing institutional resources will become focused on putting the framework mechanisms for nuclear in place.
- Undermining the political resources required for delivering renewables and demand reduction. Political resources will be harnessed for (a) taking such a big decision and (b) ensuring its movement forward, given the possibility of severe opposition.

- Undermining the financial resources available for renewables and demand reduction, whether at a government level or for private investment.
- Undermining the resources related to developing the technological requirements of a low carbon energy system.
- Undermining the development of a connected customer who takes more responsibility for his/her energy decisions, and who also may show interest in wider sustainability issues such as recycling or using public transport.

Far from nuclear power complementing a move to a low carbon energy system, it will undermine it on three levels: the technology level, the energy system level and the sustainable development level.

1. On a technology level: options are not equal and the resources and commitment required by a new programme of nuclear power would inevitably undermine commitment to renewables and demand reduction policies.

    Technology options require different financial, institutional, infrastructure and political commitments. In a world of limited resources, the scale of commitment required to deliver a new nuclear power programme would be so great that it would dwarf those available to renewables and energy efficiency. Even if additional resources were available for non-nuclear low carbon options, they would not be able to compete sufficiently with the strength of the resources behind nuclear power because of the underlying momentum of the energy system.

2. On a system level: a centralised system incorporating nuclear power is not compatible with the aim of sustainable development. A shift to a decentralised system would correspond more with the aim of sustainability. However, such a shift will require commitment and clarity of purpose because the new, low carbon technologies are effectively excluded in the UK's current electricity system. Setting up a framework to enable a new nuclear power programme would reinforce the conventional, centralised energy system and make it more difficult for a sustainable energy system to emerge.

3. On a broader level of sustainability and social change: support for a large, remote technology with inherent security problems is the antithesis of technologies which connect with people. Nuclear power therefore undermines the move to increased customer awareness of energy decisions. This in turn undermines consumer links with wider sustainability issues and the crossover benefits this brings.

Experience in Finland and the United States

Obtaining information about the requirements of a new nuclear pro-gramme is difficult because no nuclear power plants have been built and operated in a liberalised electricity system. There are 18 units actively under construction: 15 use Indian, Russian and Chinese designs. Five of them started construction prior to 1990. Detailed studies of the current and future global market implies that reports of a nuclear revival are pre-mature.[26] This paper reviews Finnish and US experience. Finland pro-vides evidence for arguing for caution against optimism and the United States provides evidence of what the nuclear industry requires for their new build.

Finland

The nuclear industry often cites the reactor in Finland as a model of nuclear construction in a liberalised market. The arguments used by the Finnish industry to justify the new construction of a 1600 MW reactor at Olkiluoto are remarkably similar to those used in the United Kingdom: the need to provide a secure supply of electricity and reduce future reliance on imports while meeting climate change commitments. Construction at Olkiluoto began in 2005 and it is due to begin operating in 2009.

However, the use of Olkiluoto as an exemplar fails to acknowledge the factors which make the new Finnish reactor unique. The reactor will be owned by TVO, a not-for profit utility owned by energy intensive Finnish industrial and power companies. TVO sells its electricity to its owners at cost.[27] This is an arrangement which is unlikely to occur in the United Kingdom. The outcome is that the owners are protected from the risks inherent in a liberalised market because they have a guaranteed market for the reactor's output.

In addition, there is some confusion about what the costs of the reactor will actually be. Reports put the costs of Olkiluoto at around 3 billion Euros, although it has also been suggested that the French vendor, Framatome, is offering it at this price as a 'loss leader'.

Areva, which is part of the consortium building the plant, will also be supported by a guarantee of over 610 million Euro from the French export credit agency, COFACE, the first time such a guarantee has been put in place for an export within the EU.[28] COFACE would reimburse this sum to Areva if the project is abandoned or if TVO is unable to hon-our the contract.

At the time of writing, the construction experience with Olkiluoto does little to inspire confidence that the project will be completed to time or cost,

Table 11.3 Measures in the 2005 Energy Policy Act 2005 ($million)[31]

	Cost ($ million)		Duration
R&D	1,432	Dept of Energy R&D programmes on new reactor technologies as well as reprocessing and transmutation	3 years
	149.7	Dept of Energy academic research and training programmes	3 years
	100	Demonstration projects for hydrogen production at existing reactors	
Construction subsidies	2,000	'Standby support' to insure the industry against delays in the construction or licensing of six new reactors. This would cover the full cost of delay for the first two reactors up to $500 million each), and 50% of the costs of delays to a further 4 reactors (up to $250 each)	First 6 reactors
	1,250	Funding of the prototype Next Generation Nuclear Plant in Idaho to produce nuclear electricity and hydrogen. Additional funds as necessary from 2016–2021	2006–2015 (2016–2021)
	~6,000[32]	Loan guarantees for up to 80% of the cost of a project. Covers 'innovative technologies', which includes advanced reactor designs	
Operating subsides		Reauthorisation of the Price Anderson Act, which caps the nuclear industry's liability in the event of an accident. The cap is set at $10 billion	Up to 2025
	5,700	Production tax credits of 1.8 cents/kWh for the first 6,000 MWh from new reactors for the first 8 years of their operation, subject to an annual limit of $215 million	2005–2025?
Back end subsidies	1,300	Changes to the tax and legal status of some decommissioning funds	

whoever bears the risks of overruns. Construction started in 2004, and is already running around a year behind schedule as a result of the complexity of the project and problems with the specification of the concrete used.[29]

United States

It is impossible to put a firm figure on the total subsidy cost for new nuclear build in the UK, as it depends both on future electricity prices and the degree of certainty required by investors. However, an indication is given by the United States government's subsidy programme, set out in the Energy Policy Act 2005. The package includes underwriting or subsidies to cover the whole lifetime of projects; a framework to insulate the industry against regulatory and legal delays; research and development funding and assistance with historic decommissioning liabilities.[30] The main measures are set out in Table 11.3.

Conclusion

The risks attached to new nuclear power build are so great that the question deserves an evidence-based, rational and transparent discussion. This chapter argues that the costs of building a nuclear power plant are inherently uncertain because of the way that construction time, load factor, cost of capital and means of waste disposal influence the final cost, and these cannot be known before the construction begins. Furthermore, this chapter argues that the risks inherent to new nuclear build are so great that it should not be attempted.

References and notes

1. PB Power (2004) *The Costs of Generating Electricity*, a Study for the Royal Academy of Engineering, http://www.raeng.org.uk/news/publications/list/reports/Cost_of_Generating_Electricity.pdf; Oxera (2005), *Plugging the Carbon Productivity Gap*, Agenda, http://www.oxera.com/cmsDocuments/Agenda_April%2005/Plugging%20the%20carbon%20productivity%20gap.pdf; The Geological Society (2005), *How to Plug the Energy Gap*, http://www.geolsoc.org.uk/template.cfm?name=PR60.
2. World Bank (1991) *Environmental Assessment Sourcebook: Volume III*. Guidelines for Environmental Assessment of Energy and Industry Projects, Technical Paper Number 154. 1991; Table 11.1 from Thomas, S. (2005) *The Economics of Nuclear Power: Analysis of Recent Studies*, Public Services International Research Unit, University of Greenwich, http://www.psiru.org/ reports/2005-09-E-Nuclear.pdf; Table 11.2 and Figure 11.1 available from BNFL (2005) submission to House of Commons Environmental Audit Committee Inquiry, Keeping the Lights on: Nuclear, Renewables and Climate Change, http://www.publications.parliament.uk/pa/cm200506/cmselect/ cmenvaud/584/584ii.pdf.

3. Load factor is the actual amount of kilowatt-hours produced in a designated period of time as opposed to the total possible kilowatt-hours that could be delivered in a designated period of time.

4. Thomas, S. (2005) *The Economics of Nuclear Power: Analysis of Recent Studies*, Public Services International Research Unit, University of Greenwich, http://www.psiru.org/reports/2005-09-E-Nuclear.pdf.

5. Thomas, S. (2005) *The Economics of Nuclear Power: Analysis of Recent Studies*, Public Services International Research Unit, University of Greenwich, http://www.psiru.org/reports/2005-09-E-Nuclear.pdf.

6. Sizewell B operating costs are the average of all 8 British Energy plants (i.e. the 7 AGRs plus Sizewell B).

7. The MIT O&M cost includes fuel.

8. 'A nuclear future fuelled by direct subsidy ruled out', *Financial Times*, 29 September 2005.

9. Mackerron, G. (2004) Nuclear power and the characteristics of 'ordinariness' – the case of UK Energy Policy, *Energy Policy* **32**, 1957—1965.

10. Standard & Poors (2005) *U.K. Security Of Supply Fears Spark Renewed Interest In Nuclear Energy*, http://www2.standardandpoors.com/servlet/Satellite?pagename=sp/Page/FixedIncomeBrowsePg&r=4&b=2&f=2&s=20&ig=&i=&l=EN&fr=4&fs=20&fig=&ft=4; UBS Investment Research (2005), *Q Series: The Future of Nuclear*, March, www.ubs.com/investmentresearch.

11. Smale, R. (2005) *UK Energy Policy*, prepared for Westminster Energy Forum, Oxford, Oxera, http://www.westminsterenergy.org/events_archive/downloads/june24/OXERA_Smale_.pdf.

12. This is very different from putting in place a policy which could provide support for nuclear power but does not clearly pick a carbon winner, for example, a carbon obligation, which would not guarantee (as far as is possible) nuclear investment.

13. Thomas, S. (2005) *The Economics of Nuclear Power: Analysis of Recent Studies*, Public Services International Research Unit, University of Greenwich, http://www.psiru.org/reports/2005-09-E-Nuclear.pdf.

14. Oxera (2005) Financing the Nuclear Option: Modelling the Costs of New Build, *Agenda*, June, http://www.oxera.com/main.aspx?id=3566.

15. British Energy (2006) Submission by British Energy Group plc to the Energy Review, April, http://www.british-energy.com/opendocument.php?did=394; BNFL (2006) BNFL Submission to Energy Consultation, March, http://www.bnfl.com/UserFiles/File/BNFL%20submission%20to%20DTI%20Energy%20Review%202006.pdf; EDF (2006) Energy Review Submission, April, http://www.edfenergy.com/core/energyreview/edfenergy-energy_review_response_main_document_v4-3.pdf.

16. European Commission (2003) Restructuring aid in favour of British Energy plc, *Official Journal of the European Union*, 31 July 2003, 2003/C 180/03, http://europa.eu.int/eur-lex/lex/LexUriServ/LexUriServ.do?uri=OJ:C:2003:180:0005:0028:EN:PDF.

17. The new reactor under construction in Finland is discussed later.

18. Mackerron, G. (1996) Nuclear power under review. In: *The British Electricity Experiment; Privatization; the Record, the Issues, the Lessons* (J. Surrey, ed.). Earthscan, London.

19. DTI and Scottish Office (1995) *The Prospects for Nuclear Power in the UK*, Cm 2860.

20. Performance and Innovation Unit (2002) *The Energy Review*, Annex 6, http://www.strategy.gov.uk/downloads/su/energy/TheEnergyReview.pdf.
21. For example, 'Russian gas row reignites nuclear debate', *The Guardian*, 2 January 2006.
22. CBI (2005) *Powering the Future; Enabling the UK Energy Market to Deliver*, Energy Brief, November 2005, http://www.cbi.org.uk/ndbs/press.nsf/ 0363c1f07c6ca12a8025671c00381cc7/e48a351cdba21f82802570b900335af9 /$FILE/Energy%20brief%20Nov%2005.pdf.
23. http://politics.guardian.co.uk/publicservices/story/0,11032,1680930,00. html
24. Poortinga, W., N.F. Pidgeon and I. Lorenzoni (2006) *Public perceptions of nuclear power, climate change and energy options in Britain*. Understanding Risk Working Paper 06-02, Norwich: Centre for Environmental Risk, http://www. tyndall.ac.uk/publications/EnergyFuturesFullReport.pdf. This finding is supported by other polls, such as Disney, H. and D. Lewis (2005), *Putting the Environment in Perspective*, The Stockholm Network, http://www.stockholm-network.org/pubs/Environment%20poll.pdf.
25. For example, BNFL (2005) submission to House of Commons Environmental Audit Committee Inquiry, Keeping the Lights on: Nuclear, Renewables and Climate Change.
26. Thomas, S. (2005) *The Economics of Nuclear Power: Analysis of Recent Studies*, Public Services International Research Unit, University of Greenwich, http:// www.psiru.org/reports/2005-09-E-Nuclear.pdf; The re-emergence of nuclear energy: an option for climate change and emerging countries? UNAM, Mexico City, April 19-20 2006 available from stephen.thomas@gre.ac.uk.
27. TVO is 57% owned by majority owned private enterprises and 43% owned by companies which are in turn majority-owned by the Government or munici-palities. International Energy Agency (2004) *Energy Policies of IEA Countries; Finland 2003 Review*, http://www.iea.org/textbase/nppdf/free/2000/ finland2003.pdf.
28. European Renewable Energies Foundation (2004), *EU Investigation Requested into Illegal Aid to Finnish Nuclear Plant*, Press Declaration 13 December, http://www.eref-europe.org/downloads/pdf/2004/EPR_Finland.pdf.
29. TVO (2006), Olkiluoto 3 into Commercial Operation in 2010, http://www.tvo.fi/926.htm.
30. ICF Consulting (2005), *2005 Energy Bill; the Impacts on Nuclear Power*, www.icfconsulting.com/Markets/Energy/Energy-Act/nuclear-power.pdf.
31. Nuclear Energy Institute (2005) *Congress Passes First Comprehensive Energy Bill in 13 Years*, http://www.nei.org/documents/Energy_Bill_2005.pdf; Public Citizen (2005) *Nuclear Giveaways in the Energy Bill Conference Report*, http:// www.citizen.org/documents/energybillnukeconfreport.pdf; Congressional Budget Office (2003) Cost Estimate S14, Energy Policy Act of 2003, http:// www.cbo.gov/showdoc.cfm?index=4206&sequence=0.
32. The $6 billion figure assumes 6 reactors with construction costs of $2.5 billion and a default rate of 50%, as estimated by the US Congressional Budget Office.

Part V Nuclear around the World

12
Nuclear Power – The European Dimension

Antony Froggatt

Introduction

The European Union has three pillars of its energy policy, environmental protection, security of supply and market liberalisation. Currently, each of these factors are under pressure due to high oil and gas prices, increasing concerns over the consequences of climate change and indications that CO_2 emissions reduction targets will not be met and a growing reliance on imported energy.

The growing pressures on traditional fossil fuel are refocusing attentions on other supply technologies, namely nuclear power and renewable energy. These are the only two widely used technologies that don't emit significant quantities of CO_2 during electricity production, but they have significantly different environmental impacts in other areas. Relating to security of supply, although the uranium fuel for nuclear power is virtually all imported, its low volume makes stockpiling more viable than for natural gas, whereas for renewable energy no fuel is required, it is an intermittent gener-ator. While the low price of gas and oil over the last decade have made both technologies relatively expensive, proponents of both technologies are keen to point out that their electricity production prices are coming down and that they are now often competitive with conventional sources.

Consequently, nuclear and renewable energies are increasingly seen as technologies that are competing to be the low carbon and indigenous energy source for Europe.

Status of nuclear sector

Somewhat surprisingly nuclear power is at the heart of the European Union institutions, as one of the founding treaties of the current EU is the Euratom Treaty. This was signed in 1957 and its preamble states, 'Recognising that nuclear energy represents an essential resource for the development and invigoration of industry'. This treaty remains intact and in force today and is likely to do so for some time to come. In fact the draft Constitutional Treaty, rejected by the Dutch and French public in 2005, proposed to retain the Euratom Treaty as a separate legal entity rather than merge it into a the new treaty as would have occurred to all other EU treaties.

The European Union has 147 nuclear reactors in operation in 13 of the 25 member states. This is 25 less than the peak of 172 in 1989. One country, Italy, abandoned an operating programme, while in the remaining 11 countries nuclear power stations were never started or completed. Currently nuclear power is used to generate around one-third of the Union's electricity. Those countries that use nuclear power and its contributions are shown in Table 12.1.

These statistics do not show the full picture of the status of nuclear power as there are variations in the level of support both in government and industry in each nuclear country. These differing levels of support can be categorised as follows:

1. Phase out is the default

In this category are countries that have phased out nuclear power and those countries that currently have life limiting decisions or conditions

Table 12.1   Contribution of nuclear power in EU member states

Country	Operating reactors	Under construction	Closed reactors	First grid connection	GWh (2005)	% Electricity
Belgium	7	0	1	1962	45,335	55.6
Czech Republic	6	0	0	1985	23,225	30.5
Finland	4	1	0	1977	22,334	32.9
France	59	0	11	1959	43,0899	78.5
Germany	17	0	19	1961	15,4612	30.1
Hungary	4	0	0	1982	13,020	37.1
Italy	0	0	4	1963	0	0
Lithuania	1	0	1	1983	10,300	69.6
Netherlands	1	0	1	1968	3,772	3.9
Slovakia	6	0	1	1972	12,335	56.1
Slovenia	1	0	0	1981	5,614	42.4
Spain	8	0	2	1968	54,895	19.6
Sweden	10	0	3	1964	70,000	46.6
United Kingdom	23	0	22	1957	75,179	19.9

upon their industry and unless these are overturned the nuclear sector will eventually be phased out. Countries included in this section are:

Belgium: In January 2003 the government passed legislation prohibiting the construction of new nuclear reactors and limiting the operating lives of existing reactors to 40 years. A change in administration has not resulted in the repeal of the legislation.

Germany: In June 2001 the red-green Government passed legislation which effectively limited the operating life of the reactors to 32 years and banned the shipment of fuel for reprocessing after July 2005. The election in September 2005 resulted in a change of administration, but not a change in position on the nuclear phase out.

Italy: Phased out nuclear power following a referendum in 1988.

Sweden: A referendum in 1979 resulted in legislation requiring the closure of all the county's nuclear reactors by 2010. This timetable has been abandoned and no definite replacement is in place. However, the two reactors at the Barseback nuclear power plant have been closed, the most recent in May 2005.

2.   Status quo with possible plant life extensions

This group of countries have an active nuclear programme, but it is not expanding. In the coming decades an increasing number of reactors would be expected to close as they reach the end of their design lives. However, the nuclear industry is seeking to both extend the lives of their existing reactor and to increase their output. These processes are potentially good for the nuclear utilities, as they can result in increased profits (as there is an increase in electricity production with relatively little capital investment) and they enable the nuclear sector to retain their share of the electricity supply industry. This group includes:

Hungary: The industry hopes to extend the operating life of the Paks nuclear power plant to 60 years.

Netherlands: The country's remaining nuclear power plant at Borssele was supposed to close in 2003; however, this government-inspired decision has been overturned and it is now planned to operate it until 2033.

Slovenia: The one reactor at Krsko is due to close in 2023 after 40 years of operation, although the government is developing plans to extend its operating life.

Spain: The government elected in 2004 announced it would gradually abandon nuclear power; however, the utilities are proposing to extend the operating lives of the reactors to 60 years.

3.   Claims for new build

The industry or government officials from the countries included in this category have called for new nuclear build. However, in themselves these statements are not sufficient to result in actual construction and therefore may or may not be of importance. Such statements have been made in the following countries.

Czech Republic: The utility CEZ is preparing proposals for the construction of two new reactors. A decision would be expected within the next few years.

Lithuania: The country is required to close its only nuclear power station, an RBMK (the same design as Chernobyl), by 2010, consequentially it is considering replacement power options, including nuclear power. One proposal is for a regional reactor to be constructed, to include utilities or governments from Poland and Estonia. A financing plan is scheduled to be developed by 2007.

4.   Key decisions imminent

In these countries the government is expected to take a decision which could lead to the start of a planning or construction process in the next 12 months. This group includes the following countries.

Slovakia: ENEL of Italy became the part owner of the Slovakian Electricity utility SE in 2005. One of the conditions for the successful privatisation bid was a requirement to complete the third and four units at the Mochovce nuclear power plant.

United Kingdom: The government is reviewing its position on nuclear power in 2006. Parts of the government, in particular the prime minister is apparently keen on constructing new nuclear power stations.

5.   Active construction programme

These countries have an active programme for new build in addition to programmes for plant life extension. This includes:

Finland: The construction of the third unit at Olkiluoto an European pressurised water reactor (EPR) designed reactor began in 2005. It was scheduled to be completed in 2009 but is already nine months behind its construction programme.

France: In 2007 the construction of an EPR is expected to begin at Flamanville. Whether this is followed by other similar reactors is unclear.

Overall, within the EU nuclear power is declining and has been for a number of years. This can be seen by the lack of orders and construction for new nuclear power plants. Between 1990 and 2010 it is likely that there will have been about 10 new reactors in current member states. However, just to retain the current nuclear capacity (i.e. to replace reactors as they are closed when they reach the end of their operational design live), around 30–40 new reactors would have had to be completed.

Despite, or perhaps because of, the lack of new construction within the EU, support for new nuclear power appears to be gradually increasing, although the majority of EU citizens remain opposed. The latest EU opinion poll says that 37% of the population is supportive of nuclear power, while 55% are opposed (the remaining 8% don't know). In only eight countries is there a majority in favour of nuclear power – Hungary, Sweden, Czech Republic, Lithuania, Finland, Slovakia, France and Netherlands. In this poll United Kingdom support was 44%. Those opposing nuclear most strongly are those countries which do not use it, with the exception of Spain where 71% are opposed (Eurobarometer, 2005).

Status of renewable sector

As noted the other low carbon option currently in deployment is renewable energy. This sector is expanding in all countries of the EU and as a result renewables contributed 14% of the EU's electricity in 2003. Figure 12.1 shows the growth of renewables of the last decade in the EU.

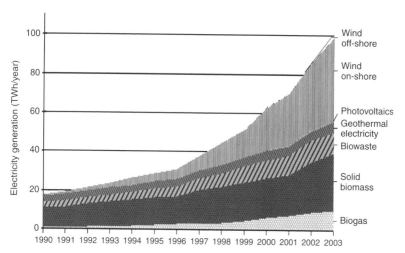

Figure 12.1   Growth of renewable electricity in member states

The European Commission forecast that growth over the next decade in renewables will be faster than any other generating source (European Commission, 2003). In order to encourage growth in 2001 the EU set a target that requires 21% of the EU's electricity from renewable energy sources by 2010 (12% of the EU's energy) (European Commission, 2001). Some technologies are fairing better than others, with wind power, now expected to nearly double a 2000 forecast and have an installed capacity of 75 GW by 2010. However, the other key sector, that of biomass has not been introduced as widely as expected. As a result and despite the growth in the wind sector, it is currently forecast by the Commission that the EU's 2010 target will not be met. The Commission expect that the output from renewable energy producers in 2010, resulting from current national policy measures, will be around 18% (European Commission, 2004a).

The failure of the EU to meet its target is mainly due to the lack of effectiveness of the support schemes in some member states. The most successful are the feed-in tariffs for wind in Denmark, Germany and Spain. In nearly half the member states the support is too low to have a significant impact on the uptake of renewables. Further obstructions to the uptake in renewables are administrative and grid barriers (European Commission, 2005).

A factor enhancing the role for renewables is the level of public support. A Eurobarometer poll published in 2006 reported that as a mechanism for reducing dependency on imported energy 48% of the population suggested that governments should focus on the development of solar power and 31% suggested wind, while only 12% nuclear power (respondents were given the opportunity to give two answers) (Eurobarometer, 2006).

Economics of new build

In December 2005 the World Nuclear Association (WNA) proudly proclaimed that nuclear power will be economically competitive even without attaching an economic weight to its 'environmental virtues or advantage of security of supply' (WNA, 2005). This is a bold statement and one which is not reflected by more independent economic commentators and analysts.

HSBC have said that the various technical and economic concerns over nuclear new build would make it a 'difficult pill to swallow for equity investors' (HSBC, 2005), while UBS stated that investment in new nuclear power in Europe was a 'potentially courageous 60-year bet on fuel prices, discount rates and promised efficiency gains' (UBS, 2005). The scepticism of the investment community is not unfounded as the industry is well known for its cost over-runs and delays. The World Bank does not fund nuclear projects, in part because suppliers 'substantially underestimate' nuclear costs (World Bank, 1992). The last reactor to be built in the United Kingdom, Sizewell B, saw costs rise from £1.6 billion to around £3.7 billion, while costs for the centrepiece of the UK nuclear industry, the Thorp reprocessing plant, rose from £300 million (at the time of its public inquiry in 1977) to £1.8 billion (when finally completed, some five years behind schedule).

More recent information on the cost reliability of new build is difficult to acquire as the lack of new construction – there are currently only 27 reactors under construction in the world – is compounded by the fact that most of these are in countries with less transparent electricity market rules (China, India and Russia have 17 of the proposed reactors). In fact globally, nuclear new build now only accounts for between 1.5% and 2.5% of all new power stations.

If a new reactor program were to be ordered in the United Kingdom it would be a Generation III reactor design. This would almost certainly be either a European pressurised water reactor (EPR), similar to that being built in Finland or the Westinghouse AP1000 presently being proposed in the United States and China, but nowhere under construction. In neither case is the design proven – there are no operating reactors to act as a reference case. An indication of potential problems to come can be seen in Finland: with construction only just over a year old the project is already nine months behind schedule. Another Generation III reactor design being built is the Advanced Boiling Water Reactor in Japan: the construction cost estimated was $1528 per kilowatt for Tokyo Electric Power Company, but when the two reactors were built the actual costs were $3236/kW for the first unit and $2800/kW for the second.

The EPR financed by Finland is also not without controversy and is the subject of an ongoing EU competition complaint. The project, highly unusually, has used export credit guarantees from two EU member states (France and Sweden) totalling €710 million to fund a project within the EU; these guarantees have been used to secure a €1.95 billion loan through the public bank of Bayerische Landesbank and other financial institutions, with a very low interest rate of 2.6%. Furthermore, the price of the turn-key construction price appears to be lower than the actual construction price: the price for a similar reactor in France is reported to be about 25% higher per kilowatt installed than the cost to Teollisuuden Voima Oy (TVO) in Finland. In its 2005 annual report Areva (the reactor vendor) admitted that the TVO reactor sale had dampened the company's profits and it had not included in the sale price the development costs of the EPR (between €1.5–2 billion).

In order to overcome the economic risks associated with nuclear new build the industry in the United Kingdom and other countries seeks financial assurances either directly from the government or via changes in the market rules. The most comprehensive support programme currently in force is in the United States which has not ordered and completed a nuclear reactor for 30 years, but despite this has a very successful lobby in Washington. In the 2005 energy act it managed to have passed legislation that will effectively give around $12 billion in further subsidies to the nuclear industry, as it tries to build reactors once again. These assistance programmes include a production tax credit for the first six reactors, government guarantees against licensing delays, addition research and development funding and assistance on decommissioning (FOE, 2005).

It is unlikely that such a broad ranging support programme would be acceptable in the United Kingdom or the EU as it would almost certainly be incompatible with the current EU's State Aid rules. However, a number of mechanisms are being considered, these include reducing the time needed for the licensing process, greater government responsibility for waste management and even guarantees on electricity prices.

New energy policy for Europe

At the informal EU Summit in Hampton Court in October 2005, Tony Blair put forward a paper written by Dieter Helm of New College Oxford, called 'European energy policy' (Helm, 2005). This publicly opened a discussion on the development of a new energy policy for Europe. Over the New Year a payment dispute between Russia and Ukraine led to the cutting of a natural gas supplied to Ukraine, with subsequent reports that Western European gas supplies would be affected. This was later shown not to be the case, but the Ukrainian–Russian gas dispute led further calls for common European action on energy policy. As a result in March 2006 the European Commission published a Green Paper on a European Strategy for Sustainable, Competitive and Secure Energy. Until now there has been no common EU energy policy as this was deemed to be the sole competence of member states. Common EU rule and policies are already in place in different energy sectors, such as market liberalisation rules or emissions trading schemes, but there is no overriding energy policy. The Green Paper lists a number of priority areas for action: completing the internal European electricity and gas markets; security of supply; solidarity between member states; achievement of a more sustainable, efficient and diverse energy mix; development of an integrated approach to tackling climate change; encouraging innovation; and creating a coherent external energy policy (European Commission, 2006). The future of the EU's energy policy the ideas in the Green paper will be developed further and discussed at the March 2007 Summit, held under the German Presidency, where an action plan on a common Energy policy will be adopted.

On the issue of choice of generating capacity the Green Paper is initially clear as it says that 'each member state and energy company chooses its own energy mix'. However it later suggests that the EU could adopt an overall strategic framework for energy policy which might include 'a minimum level of the overall EU energy mix originating from secure and low-carbon energy sources'. The 'low carbon' terminology was also proposed by the European Parliament, as in September 2005 when it called for CO_2 neutral and very low emission technologies to supply 60% of the EU's electricity demand by 2020 (European Parliament, 2005). As the EU already has targets for the use of renewable energy the introduction of additional targets for low carbon technologies would currently de facto lead to nuclear targets.

The EU has defined and confirmed on a number of occasions, an indicative long-term global temperature target of not more than 2°C above pre-industrial levels (European Council, 2004). In order to achieve this, the world will have to reduce emissions far more substantially than the 8% by 2008–2012 mandated (for the EU) by the Kyoto Protocol. This will require urgent action to reduce emissions across all sectors and in particular on the short term from the power sector, where generation switching and energy efficiency can achieve rapid results. How much of this reduction will come from the introduction of renewables and how much from nuclear power is currently under debate in a number of countries and in the EU institutions. This, along with the low carbon language of the Green Paper on energy policy encourages nuclear power and renewables to compete to be the CO_2 reduction technology.

Although both nuclear and renewable energy technologies can be classified as CO_2 neutral or very low emitters, for virtually all environmentalists, with one or two exceptions, like James Lovelock, this is where the environmental acceptability ends. Nuclear power generate wastes across its whole fuel cycle, from uranium mining, fuel enrichment and fabrications, operational discharges, spent fuel management and finally the decommissioning of facilities. In the United Kingdom the expected cost of this is expected to exceed £100 billion. The transportation and reprocessing of nuclear fuel, a practice undertaken by some European countries, further exacerbates the environmental problems. All of these, along with the threat of a nuclear accident and proliferation of nuclear materials must be compared to the localised environmental impact of most renewable energy generators. This is why European environmental groups pushed so hard for nuclear power to be excluded lists of clean technologies, such as the Clean Development Mechanism of the Kyoto Protocol.

Analysis from the United States suggests that nuclear power is the least effective climate stabilising option available and that its deployment may actually retard carbon displacement, as its investment in nuclear will be significantly less effective than renewable energy in reducing CO_2 emissions (Lovins, 2005). The UK Government's Sustainable Development Commission reported in March 2006 that the problems of nuclear power, namely its high cost, there is no long-term solution for radioactive waste management, it is inflexible and will lock an energy system into a centralised distribution system, it will undermine energy efficiency and potentially leads to greater international security risks, outweigh any low carbon advantages that it might bring (SDC, 2006).

Conclusions

The role of Brussels in energy policy will increase over the coming years. This is in response to a belief that common European action is needed on the traditional areas of safeguard supplies, to reduce the environmental impacts of the energy sector and to ensure the smooth running of the energy markets. However, there will also be a growing issue of energy and foreign policy.

In this era of increasing influence of Europe in national energy policy, it is likely that based on past experience nuclear power will potentially gain, as it has historically been well supported by the European treaties and policies. However, nuclear power is not universally supported with 12 of the EU's 25 member states not operating nuclear power plants and only one reactor under construction. Furthermore, the prevailing public view is that the nuclear power should not be supported, rather government attention should be focused towards the further development of renewable energy.

Therefore politicians face some key questions. Nuclear power is currently declining, as the decades of lack of new build have left an ageing reactor fleet. The United Kingdom, with the oldest reactors, will be the first to begin an industrial-led phase out. Already the EU has just 147 reactors in operation, 25 less than the historic peak of 172 in 1989. Just to maintain the output from the nuclear sector will require the completion of around 3 reactors each year, but it is likely that only two reactors will be completed in the next decade. Therefore the decline in nuclear will continue for at least another 15 years and almost certainly longer. To build a significant new nuclear power programme will require major changes to the ways in which nuclear power is financed, which would almost certainly require changes to the energy market.

Such major changes would need major political and public support, which does not exist for nuclear power, but it does for the other low CO_2 emitting technology, renewable energy. Given the urgency and importance of some of the problems facing the energy sector priority should be given to technologies that have wider public and political support.

Unfortunately, the most likely scenario is that some politicians will support a programme of new build for nuclear power. However, for technical or economic reasons the programme will not get off the ground, but in the meantime political and financial support will be diverted away from renewables and energy efficiency and therefore the EU's climate and security of supply programme will be further damaged.

References

Eurobarometer (2005) Special Eurobarometer, *Radioactive Waste*, European Commission, June.

Eurobarometer (2006) Special Eurobarometer, *Attitudes Towards Energy*, European Commission, January.

European Commission (2001) Directive 2001/77/EC on *the promotion of electricity produced from renewable energy sources* (OJ L283/33, 27.10.2001).

European Commission (2003) *European Energy and Transport Trends to 2030*. European Commission ISBN 92-894-4444-4.

European Commission (2004a) Communication from the Commission to the Council and the European Parliament: The share of renewable energy in the EU Commission Report in accordance with Article 3 of Directive 2001/77/EC, evaluation of the effect of legislative instruments and other Community policies on the development of the contribution of renewable energy sources in the EU and proposals for concrete actions. COM (2004) 366 final, 26th May.

European Commission (2005) Communication from the Commission. The support of electricity from renewable energy sources. COM (2005) 627, 7th December.

European Commission (2006) Green Paper, a European strategy for sustainable, competitive and secure energy. COM (2006) 105 final, 8th March.

European Council (2004) *Spring European Council 2004, 25–26 March 2004*. Doc 7631/04.

European Parliament (2005) *Motion for a European Parliament Resolution on the share of renewable energy in the EU and proposals for concrete actions*. 2004/2153(INI).

FOE (2005) Energy Bill: Billions More in Taxpayers Handouts to the Failed Nuclear Industry. Friends of the Earth, *Public Citizen*, Earth Track, US PRIGG.

Helm, D. (2005) European Energy Policy, Securing Supplies and Meeting the Challenge of Climate Change, October 2005 http://www.fco.gov.uk/Files/kfile/PN%20papers_%20energy.pdf.

HSBC (2005) Nuclear Clean Up Cost: 2020 Fission. A financial Analysis of Nuclear Decommissioning Costs. HSBC Global, September.

Lovins, A. (2005) *Nuclear Power Economics and Climate-Protection Potential.* The Rocky Mountain Institute. September 2005.

SDC (2006) *The Role of Nuclear Power in a Low Carbon Economy.* Sustainable Development Commission, March.

UBS (2005) *More a Question of Politics than Economics.* Q Series: The Future of Nuclear, UBS Investment Research, March 2005.

World Bank (1992) *Environmental Assessment Sourcebook, Volume III Guidelines for Environmental Assessment of Energy and Industry Projects.* World Bank Technical Paper Number 154, 1992.

WNA (2005) *The New Economics of Nuclear Power.* World Nuclear Association, December.

13

Non-nuclear Sustainable Energy Futures for Germany and the UK

Godfrey Boyle

Introduction

It is sometimes argued in Britain and elsewhere that nuclear power must be an essential element in the electricity generating mix of large, developed countries if they are to make major cuts in CO_2 emissions by mid-century. For example, the UK Prime Minster, Tony Blair, addressing the Labour Party Conference on September 27 2005, urged the nations of the world, in response to the challenge of climate change, 'to develop together the technology that allows prosperous nations to adapt and emerging ones to grow sustainably; and that means an assessment of all options, including civil nuclear power'.

But the experience of Germany to date, and her ambitions plans for the future, suggest otherwise. Germany is a larger and wealthier nation than Britain, with higher electricity consumption and a higher proportion of nuclear power generation, but with poorer fossil and renewable energy resources. Yet she is phasing out nuclear energy by 2020, phasing in renewable energy many times faster than the United Kingdom and has detailed plans to cut carbon emissions by 80% by 2050.

In an interview on 21 October 2005, Germany's outgoing environment minister Jurgen Trittin summarised his country's contrasting attitudes to nuclear power and renewable energy as follows:

> The safety risks associated with nuclear power have in no way decreased in recent years – in particular with regard to the threat of terrorism, they have in fact increased dramatically. And as far as the long-term management of radioactive wastes is concerned, we are fundamentally no wiser than we were 30 years ago. The use of nuclear power is and will remain a global risk, especially for future generations. (…)

> In contrast, renewable energies are essential to solving pressing issues
> for the future. (…) With the further rapid expansion of wind and
> hydropower, solar power, the use of biomass and geothermal power –
> we can create an alternative to a nuclear and fossil fuel energy supply in
> a step-by-step process. We have already made good progress. In 2004,
> 9.8% of electricity in Germany came from renewable sources. Ten years
> ago, the figure was not even half this. And this trend is set to continue.
> Our goal in Germany is to provide at least 20% of electricity from
> renewable energies in the year 2020. And by the middle of the century,
> we want to cover about 50% of our total energy consumption with
> renewable sources. (…) (Trittin, 2005)

Germany's energy and environment policies over the past decade demonstrate the practicality of an energy strategy involving the very rapid
deployment of renewable energy sources, coupled with increasing
improvements in energy efficiency, and decreasing reliance on nuclear
power for electricity generation.

Moreover, the success of these policies to date lends credibility to the
German environment ministry's detailed long-term feasibility studies of
future energy supply and demand over the decades to 2050, which suggest
that renewable energy could by then be contributing around half of the
country's energy needs, with CO_2 emissions cut by 80% below 1990 levels.

Germany and the UK: Current sustainable energy and climate policies

As Table 13.1 shows, Britain and Germany have similar gross domestic
product (GDP) per capita and population density. But Germany has a higher
installed nuclear power capacity than the United Kingdom, and this supplies a higher proportion of the nation's electricity. In 2003–2004, Britain's
renewable energy sources contributed 1.3% of the country's primary
energy and 3.5% of its electricity. By contrast, renewables in Germany
contributed some 3% of primary energy and 7.9% of electricity in 2003 –
more than twice as much as in the United Kingdom. And by the end of
2004, as Jurgen Trittin points out in the interview quoted above, the
contribution of renewables to electricity generation had risen to 9.8%.

The rate of growth in Germany's renewable energy supplies in recent
years has been astonishing: between 1998 and 2003 the contribution of
biomass energy doubled, wind power capacity quadrupled and the number of solar photovoltaic roofs increased six-fold.

By 2003–2004, Germany's installed wind and solar photovoltaic capac-
ities were respectively 19 and 70 times more than those of the United
Kingdom, as Table 13.1 shows.

Table 13.1 Germany and the UK: Selected comparative data (2003–2004)

	Germany	United Kingdom
GDP (2003)	$2,270 billion	$1,666 billion
GDP per person	$27,550	$27,630
Population	82.4 million	60.3 million
Land area	349,000 sq km	242,000 sq km
Population density (persons per hectare)	2.4	2.5
Annual electricity demand (TWh) (2003) (1 Terawatt-hour (TWh) =1 billion kWh)	506 TWh	338 TWh
Annual electricity use per person, kWh (kilowatt-hours)	6140 kWh	5578 kWh
Percentage of electricity from nuclear (2003)	28.8%	22.7%
Percentage of electricity from renewables (2003)	7.9%	3.5%
Percentage of primary energy from renewables (2003)	3%	1.3%
Capacity of wind power installed (2004)	16,600 megawatts	880 megawatts
Number of photovoltaic roofs and capacity (2003)	>100,000 410 megawatts	<1000 5.9 megawatts

Although premium prices are paid for renewable power under Germany's Renewable Energy Sources Act, the additional costs are modest – one Euro per month per household – and are added to electricity bills, not paid through taxes. The prices are different for each technology and the subsidy system is quite sophisticated. For example, each year the price paid for electricity from new photovoltaic installations falls by 5%, giving solar manufacturers a strong incentive to reduce prices as the size of their market expands. But the premium prices are guaranteed for 20 years, giving confidence to investors.

Alongside measures to promote renewables, Germany has also been strongly encouraging more efficient use of energy, for example through incentives for combined heat and power generation and increasingly stringent regulations on the energy performance of buildings.

UK and German energy and greenhouse gas reduction plans to 2010–2020

So how do Germany's and Britain's plans for the rest of this decade and beyond compare?

The UK government's 2003 White Paper on energy (DTI, 2003) emphasised the role of renewables, combined with heat and power and energy efficiency, in enabling the UK to meet its Kyoto treaty commitment to cut greenhouse gas emissions (mainly carbon dioxide, but including other gases) by 12.5% by 2012. No new nuclear power stations would be built, though the option of doing so in future was left open.

By the end of 2004, the United Kingdom had reached its Kyoto target, though there are concerns that emissions may rise again in future years. By means of the renewables obligation, coupled with other measures such as the climate change levy, the government plans to increase the proportion of renewable electricity to 10% by 2010 and to 20% by 2020. It has also pledged to go beyond Kyoto and cut the emissions of the principal greenhouse gas, CO_2, by 20% by 2010.

Germany's renewable electricity targets are similar: 12.5% by 2010 and 20% by 2020. But by 2010 it also aims to achieve a 10% contribution of renewables to *primary* energy. Germany's Kyoto target is for a 21% cut in greenhouse gas emissions. By 2004, it had reached 19%.

Long-term German energy scenarios to 2050

Germany's ambitious plans for the rest of this century are described in detail in the Environment Ministry's 2004 report *Ecologically-Optimised Extension of Renewable Energy Utilisation in Germany* (BMU, 2004). By 2050, the report envisages primary energy use falling to around half the current level, despite continuing economic growth and rising prosperity, due to major improvements in energy efficiency (Figure 13.1) and increasing use of combined heat and power plants (Figure 13.2). By then, renewables should be supplying 65% of the nation's electricity, 45% of its heat and 30% of its transport fuel (Figure 13.3). Nuclear power will have been phased out three decades ago and fossil fuel use reduced to around 20% of current levels. This 'ecologically-optimised' energy system should allow Germany to achieve an 80% cut in greenhouse gas emissions, making a major contribution to international efforts to mitigate climate change, and setting an example to other wealthy nations.

As shown in Figure 13.1 primary energy consumption/GDP in 1970 was 9.93 GJ per thousand Euros (in 2000 money terms). Electricity consumption/GDP in 1970 was 0.787 GJ per thousand Euros (in 2000 money terms). The energy intensity index for 1970 was 100. The historic rate of improvement in primary energy intensity and electricity intensity in Germany was approximately 1% per annum between the mid-1980s and 2003. Improved energy efficiency measures envisaged in the 'ecologically optimised' extension scenarios are estimated to raise the rate of energy and electricity intensity improvement to around 2% per annum in the period to 2050.

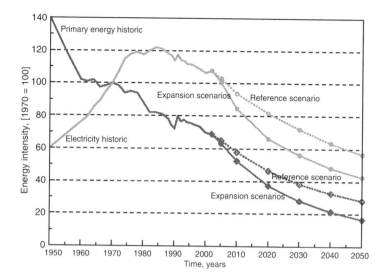

Figure 13.1 Energy intensity curves (primary energy consumption/GDP; electricity/GDP) for Germany since 1950 and in the BMU's reference scenario and extension scenarios to 2050 (*Source*: BMU, 2004)

Long-term UK energy scenarios to 2050

The RCEP long-term energy scenarios

In 2000, the UK's Royal Commission on Environmental Pollution (RCEP) published four long-term energy scenarios for the UK, each aimed at achieving a 60% cut in carbon emissions by 2050. The scenarios are summarised in Box 1, and in more detail in Tables 13.2 and 13.3.

The key parameters for these four scenarios are as follows:

The four scenarios differ (a) in the extent to which overall energy demand is reduced, (b) in the contribution made by renewable sources and (c) according to whether baseload electricity from large plants is generated from nuclear power or from fossil fuels with carbon capture and sequestration.

Two of the scenarios involve an expansion of nuclear power; the other two envisage a low or zero contribution. All of the scenarios envisage a major expansion in renewable energy use, and in all of them the fossil fuel contribution is around 40%, much reduced from 1990 levels but still a major proportion of energy use.

Table 13.3 summarises in more detail the outputs of the various energy sources envisaged in each of the four scenarios for 2050.

188

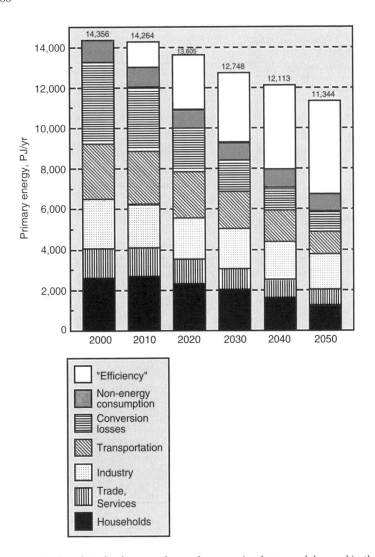

Figure 13.2 Trends in final energy demand, conversion losses and demand in the transportation, industry, trade & services and household sectors and the resulting net primary energy envisaged in the BMU 'extension scenarios' for Germany 2000–2050. 'Efficiency' in the figure is the additional reduction, compared with the reference scenario, that can be achieved by means of increased energy efficiency and increased combined heat and power generation (*Source*: BMU, 2004)

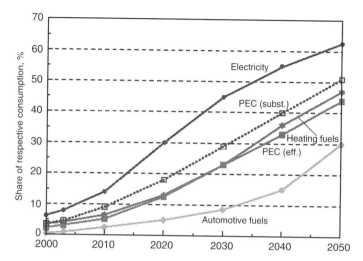

Figure 13.3   Growth of percentage of renewable energy contribution to electricity, heating, transport fuels and primary energy consumption (PEC) envisaged in the extension scenarios for Germany, 2000–2050 (PEC is calculated here using both the 'efficiency' method and the 'substitution' method)

Box 1: Four Scenarios for 2050

Four scenarios were constructed to illustrate the options available for balancing demand and supply for energy in the middle of the 21st century if the UK has to reduce carbon dioxide emissions from the burning of fossil fuels by 60%:

scenario 1: no increase on 1998 demand, combination of renewables and *either* nuclear power stations *or* large fossil fuel power stations at which carbon dioxide is recovered and disposed of

scenario 2: demand reductions, renewables (no nuclear power stations or routine use of large fossil fuel power stations)

scenario 3: demand reductions, combination of renewables and *either* nuclear power stations *or* large fossil fuel power stations at which carbon dioxide is recovered and disposed of

scenario 4: very large demand reductions, renewables (no nuclear power stations or routine use of large fossil fuel power stations).

Table 13.2   RCEP scenarios

	Scenario 1	Scenario 2	Scenario 3	Scenario 4
Percentage reduction in 1997 carbon dioxide emissions	57	60	60	60
DEMAND (%) Reduction from 1998 final consumption				
Low-grade heat	0	50	50	66
High-grade heat	0	25	25	33
Electricity	0	25	25	33
Transport	0	25	25	33
Total	**0**	**36**	**36**	**47**
SUPPLY (GW) Annual average rate				
Fossil fuels	106	106	106	106
Intermittent renewables	34	26	16	16
Other renewables	19	19	9	4
Baseload stations (either nuclear or fossil fuel with carbon dioxide recovery)	52	0	19	0

Source: RCEP, 2000.

The RCEP scenarios envisage a large proportion of electricity generation in future being based on smaller-scale plants, to avoid the waste currently entailed in large-scale generation remote from centres capable of using the waste heat (Table 13.4). Such plants could use combined cycle gas turbines, which are economic on relatively small scales, internal combustion engines or fuel cells, with the plants sited to facilitate the use of waste heat wherever possible. Such plants would not, however, be used for 'baseload' electricity generation. That role would be performed by either nuclear power stations or large fossil fuelled power stations fitted with carbon capture and sequestration systems. It would probably not be practicable to utilise most of the waste heat from such large stations.

The smaller electricity generating plant would operate alongside bio-fuelled combined heat and power (CHP) plant but would be mainly fossil-fuelled. They would only be used relatively infrequently to meet peak demands, so their carbon emissions on an annual basis would be relatively low. When electricity demand increased towards peak levels, the electricity output of the biofuelled CHP plant would be increased in response; then, on the rare occasions when output was still insufficient, the additional

Table 13.3 Outputs from UK energy sources (annual average rate) at present and in 2050 in the four RCEP scenarios

Source	Present output (GW)	Output in Scenario 1 (GW)	Output in Scenario 2 (GW)	Output in Scenario 3 (GW)	Output in Scenario 4 (GW)
Onshore wind	0.10	6.5	3.3	0.2	3.3
Offshore wind		11.4	11.4	11.4	5.7
Solar PV		10.0	5.0	0.5	0.5
Wave		3.75	3.75	3.75	3.75
Tidal stream		0.25	0.25	0.25	0.25
Tidal barrage		2.2	2.2	0.0	2.2
Total intermittent renewable sources		34.1	25.9	16.1	15.7
Hydro existing	0.59	0.59	0.59	0.59	0.59
Hydro new small scale	0.02	0.3	0.3	0.3	0.2
Energy crops		10.2	10.2	1.8	1.8
Agricultural forestry waste	0.04	5.7	5.7	5.7	1.2
Municipal solid waste	0.15	1.9	1.9	0.0	0.0
Total renewable sources		52.8	44.6	24.5	19.5
Nuclear power*	11.4	52	0	19	0
Contributions from fossil fuels	266	106	106	106	106

*Alternatively, the same amount of energy might be provided by fossil fuel baseload stations at which carbon dioxide is recovered and disposed of.
Source: RCEP, 2000.

local fossil-fuelled generating plant would be called into service to meet peak demands, with its 'waste' heat also being supplied to the district heating networks. For the next few decades, infrequently-used 'standby' electricity supplies could come from existing fossil-fuelled power plants; but in the longer term new fossil-fuelled backup power plants would have to be constructed (RCEP, 2000).

The two RCEP scenarios that do not involve contributions from nuclear power illustrate the validity of an alternative approach to UK energy policy, similar to that of Germany, which would rely on an increased contribution from renewables and greater reductions in energy demand, coupled with a substantial role for combined heat and power and carbon capture and sequestration (CCS).

Table 13.4 Number of generating plants required in 2050 in the four RCEP scenarios

	Scenario 1	Scenario 2	Scenario 3	Scenario 4
Large onshore wind farms (100 turbines each)	50	25	2	25
Small onshore wind farms (10 turbines each)	510	250	16	252
Large offshore wind farms (100 turbines each)	180	177	180	88
Photovoltaic roof installations (average peak output 4 kW)	15 million	7.5 million	0.75 million	0.75 million
Wave power units (1 MW capacity)	7,500	7,500	7,500	7,500
Tidal stream turbines (1 MW capacity)	500	500	500	500
Tidal barrage	1	1	1	1
New small scale hydro	4,500	4,500	4,500	2,200
CHP plants fuelled by energy crops (capacity 1–10 MW)	290–2,900	290–2,900	42–420	42–420
CHP plants fuelled by agricultural and forestry wastes (capacity 0.5–10 MW)	53–1,050	53–1,050	53–1,050	34–688
CHP plants fuelled by municipal solid waste (capacity 8–60 MW)	3–20	3–20	0	0
Baseload plants: *either* nuclear *or* fossil fuel with carbon dioxide recovery and disposal (capacity 1,200 MW)	46	0	19	0
Domestic (micro) CHP units using gas (2 kW)	0	1.7 million	1.8 million	2.4 million
Fossil fuel plants to back up intermittent renewables (capacity 40 MW)	1,000	760	475	460
Fossil fuel plants for meeting peak electricity demand (capacity 400 MW)	120	70	65	55

Source: RCEP, 2000.

The RCEP report is not strictly an official government document, but its target of a 60% reduction in carbon emissions by 2050 has been accepted by the UK government, even though the detailed prescriptions described in its energy scenarios are still being discussed and have only been implemented to a limited extent.

Energy scenarios from DTI/Future Energy Solutions

In 2003, as background to the UK Department of Trade and Industry's (DTI's) 2003 Energy White Paper *Our Energy Future – A Low Carbon Economy*, Future Energy Solutions published the results of a detailed study, commissioned by the DTI, of *Options for a Low Carbon Future*. This included the results of modelling a range of long-term UK future energy scenarios using the MARKAL model. Examples of the FES scenarios are shown in Figures 13.4, 13.5 and 13.6. Their report concluded, inter alia, that:

> Too often the debate on reducing carbon emissions has been polarised between the advocacy of nuclear power on the one hand and renewable energy on the other. This is unfortunate since it obscures what is a rich range of options for achieving a low carbon future (...).

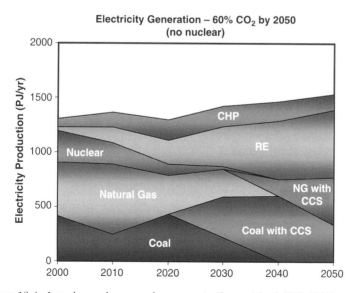

Figure 13.4 Low demand non-nuclear scenario (*Source*: Marsh/FES, 2005)

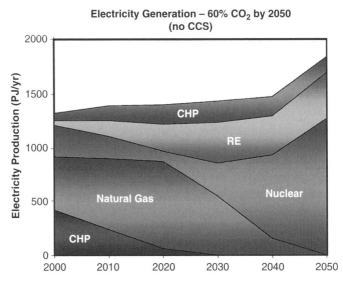

Figure 13.5 Medium demand nuclear expansion scenario (*Source*: Marsh/FES, 2005)

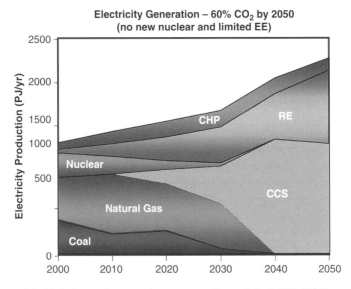

Figure 13.6 High demand non-nuclear scenario (*Source*: Marsh/FES, 2005)

The present study has confirmed the Royal Commission on Environmental Pollution's (RCEP) conclusion that the replacement of current nuclear power stations by new nuclear stations, and an expansion of nuclear power, could help the UK reduce its CO_2 emissions by 60% or more by 2050. It has also shown that similar amounts of energy could be delivered by renewable energy sources. (…)
In sensitivity studies with the GS (Global Sustainability) scenario the introduction of a 'no nuclear' constraint did not prevent the attainment of the carbon abatement targets, and only increased costs with the maximum 70% reduction. The effect of reducing nuclear power is to put more onus on renewable energy, biomass and carbon sequestration, and so forth. A low carbon future is technically and economically feasible without nuclear power.

In the scenario described in Figure 13.4, electricity demand by 2050 increases marginally over 2000 levels, nuclear energy is phased out by 2030 and demand is met by a large contribution from renewables, a similar contribution from fossil fuels (coal and natural gas) with CCS, and a moderate contribution from CHP.

In the scenario described in Figure 13.5, electricity demand by 2050 increases by ca. 40% over 2000 levels. No carbon capture and storage are used, coal use is phased out by ca.2030 and natural gas use by 2050. There is an initial decline in nuclear generating capacity, followed by major expansion from 2020 to around two-third of electricity supply by 2050, plus a moderate contribution from renewables and a small contribution from CHP.

In the scenario described in Figure 13.6, electricity demand in 2050 is ca. 75% higher than in 2000 due to limited implementation of energy efficiency measures, and is met by a combination of fossil fuels (coal and natural gas) with carbon capture and storage, renewables and CHP. Nuclear power is phased out by 2030.

Conclusions

Germany's success over the past decade in deploying renewable energy sources and improving energy efficiency, together with the country's detailed plans to phase out nuclear power by 2020 and to develop renewables to provide 50% of primary energy by 2050, demonstrate the feasibility of non-nuclear sustainable energy strategy in a major industrialised country.

The feasibility of such an approach in Britain is illustrated in Scenarios 2 and 4 from the RCEP. These envisage substantial reductions in primary energy demand, no nuclear power contribution, and major increases in supply from renewables and from fossil fuels with carbon capture and storage. Subsequent detailed modelling studies by Future Energy Solutions for the DTI explicitly confirm that 'a low-carbon future is technically and economically feasible without nuclear power'.

Postscript on recent political developments in Germany

Recent political changes in Germany have resulted in a continuation of the previous government's energy policies, including continuing support for renewable energy and energy efficiency, and the phasing out of nuclear power by 2020. In any case it was very unlikely that any new German government would have wished to undermine the future of its now very successful sustainable energy industry, which in 2004 employed 150,000 people and had a turnover of 11.6 billion Euros (Trittin, 2005). Indeed the new coalition Government further reinforced its predecessor's sustainable energy policies in December 2005 by allocating new funding of 1.5 billion Euros per year to upgrade the pre-1978 housing stock to higher energy standards.

References

Blair, A. (2005) Speech to Labour Party Conference, 27 September 2005, From Labour Party web site http://www.labour.org.uk/conference2005.

Department of Trade and Industry (DTI) (2003) *Our Energy Future: creating a low-carbon economy.* The Stationery Office, London. 142 pp.

Department of Trade and Industry (DTI) (2003) *Options for a Low Carbon Future (DTI Economics Paper No 4)* Department of Trade and Industry, London. 178 pp.

Federal Ministry for the Environment, Nature Conservation and Nuclear Safety (BMU) (2004) *Ecologically Optimised Extension of Renewable Energy Utilisation in Germany – Summary.* BMU, Berlin. 48 pp.

Marsh, G. (2005) Presentation on behalf of Future Energy Solutions to Parliamentary Renewable and Sustainable Energy Group, Westminster, 26 January 2005.

Royal Commission on Environmental Pollution (RCEP) (2000) *Energy – The Changing Climate.* The Stationery Office, London. 292 pp.

Trittin, J. (2005) Interview on BBC web site, 21 October 2005 http://news.bbc.co.uk/1/hi/sci/tech/4357238.stm.

14
Nuclear Power around the World

Stephen W. Kidd

Introduction

Opinions about nuclear power in the United Kingdom are very much coloured by the history of nuclear in this country. External events have also been important, notably the accidents at Three Mile Island and Chernobyl, plus the concerns over international nuclear proliferation surrounding countries such as Iran, but there are several domestic features that are crucial. The experience of nuclear in other parts of the world has, in many cases, been much happier and explodes many of the myths about nuclear which have built up over the years. The current increase in interest in nuclear is leading to much debate about enhanced international co-operation, which should support a revival in nuclear plant construction in many countries, including some which have never before had nuclear plants. The arguments used against expanding nuclear power are increasingly looking very weak, but renewed growth will require a great deal of political will combined with considerable energy and innovation from the industry and its backers.

Nuclear in the UK

Commercial nuclear power in the United Kingdom grew directly out of the nuclear weapons programme of the late 1940s and 1950s. The first generation of reactors were essentially 'dual use' in that they were used to supply plutonium for bombs at the same time as generating electricity. The early enrichment facilities likewise produced highly enriched uranium (HEU) for bombs as well as low enriched uranium (LEU) for the next generation of reactors. It has been hard for the civil sector to throw off these military origins and much of the secrecy and general suspicion that surrounds it undoubtedly have their origins here. It would also be fair to say that those responsible for creating a more positive image for the civil side did a very poor job. Indeed, it could almost be said that nobody really tried at all – the public was expected to be extremely grateful for the 'gift' of nuclear power, without having it explained to them properly.

The image of civil nuclear power in the United Kingdom has also undoubtedly been soured by the experience of previous reactor programmes. Political interference and inconsistencies in policy-making led to the adoption of a poor reactor choice for the United Kingdom. Persisting with gas-cooled reactors as a nationalistic option when the world was adopting light water reactors contributed to an inferior economic performance. This was worsened by the time it took to build the reactors – up to 20 years in some cases. If nuclear power is to be economic, reactors have to be brought on-line quickly and this has never happened in the United Kingdom. This has led to the general assumption that nuclear power will always be a non-economic choice and requires government subsidies to survive.

The experience of waste management and decommissioning has also been unhelpful. The reprocessing of used nuclear fuel has been a major part of the UK's strategy, as it has in several other countries. Yet this programme, centred at Sellafield, has also been subject to delays, technical problems and attendant cost overruns. Beyond this, the failure to agree on public policy for the disposal of radioactive wastes has been a barrier to explaining that the industry has sound technical solutions available, which can be readily financed as part of the nuclear fuel cost. Finally, the huge sums now required to decommission nuclear sites in the United Kingdom have attracted a lot of attention. The overwhelming majority of this is attributable to the early days of the military-based programme, when there was a great deal of experimentation taking place, but there is nevertheless an assumption that the civil industry is incapable of financing its decommissioning liabilities and must seek recourse to the public purse.

Nuclear worldwide

It is often believed that nuclear power has been in decline throughout the world for many years. There have been relatively few new reactor start-ups in the Western world for many years, while the numbers of gas-fired plants have spiralled. Yet the nuclear share of world electricity generation has remained remarkably constant at around 16% since the late

1980s, a period of almost 20 years. Nuclear power generation has kept up with growing overall electricity supply. How has this happened?

The answer lies partly in the shift of the centre of gravity in nuclear towards Asia. Although North America and Western Europe have almost stopped commissioning new reactors, the nuclear programmes in Japan, Korea and more recently China and India have been pushing ahead. Yet the number of reactors in operation (around 440 today) is little changed from the late 1980s, with a few going out of service each year and a similar number of new ones starting up. The new reactors, however, are much larger than those shutting down, so world nuclear generating capacity has been increasing slowly but steadily. Another very important factor in this has been capacity up-rates at existing reactors. Plant operators have found this to be a highly economic way of gaining more power – indeed, up-rates in the United States have already added the generating capacity of several new reactors.

The most important factor in increasing nuclear generation has, however, been the improved operating performance of existing reactors, many of them 20 or 30 years old. Some of this can be attributed to electricity liberalisation in many markets, but whatever the cause, nuclear plants in many countries are now operating 90% of the time, whereas before 70% was regarded as an adequate performance. So despite only slow growth in world nuclear generating capacity, electricity production has risen much more quickly. This has transformed the economics of nuclear power as reactors are now generating power at costs below those of rival technologies such as coal, gas and oil. Even taking into account the heavy investment costs of new nuclear plants, building new ones can make sound economic sense if they can be built quickly and also operate at high capacity factors for many years. This is particularly so at a time when prices of the fossil fuels have been rising sharply. Improved economic performance has also led to applications for operating license extensions for nuclear plants, such that many are now expected to operate for up to 60 years. So the number of plant closures has been limited, although over half of the world's reactors are now 22 years or more from their start-up.

Particular countries

The experience of nuclear power in France contrasts considerably with that of the United Kingdom. France took a long-term strategic decision in the aftermath of the first oil shock in 1973 to use nuclear to address its weaknesses in energy security, given its poor access to resources. This has been backed by a consistency of government policy-making, sound planning and following up with sufficient financial and other resources to allow nuclear to supply almost 80% of France's electricity today. By adopting a good reactor design, then replicating this in great numbers through three succeeding generations of increasing power ratings, the country has achieved relatively cheap and secure electricity supplies, also with very little environmental impact. Waste management and decommissioning have also been handled very much better than in the United Kingdom.

Elsewhere in Europe, the picture is more mixed. Germany and Sweden have both had generally successful nuclear programmes but have embarked on slow nuclear phase-out policies, under pressure from minority government parties of a green hue. Italy turned away from nuclear after early reactors were completed, but now feels vulnerable to fossil fuel prices and power imports. The general picture everywhere is much the same in that the current operating reactors are doing very well and earning significant profits for their owners at a time of generally high electricity prices. The excellent historical performance of the Finnish reactors (two of which are of Soviet design) has undoubtedly been a major factor behind the decision to construct a fifth reactor.

In Eastern Europe and the former Soviet Union, much of the attention has had to be focused on improving the safety of flawed reactor designs in the aftermath of the Chernobyl accident and also in completing reactors under construction at the time of the fall of the Soviet Union. This period is now largely over, with both safety and economic operation greatly enhanced, and a new era where new nuclear build is a serious possibility has been entered.

To some extent, nuclear power in North America has suffered from similar ills as the British programme. The military legacy has been difficult to shake off, while construction delays and poor operating performances (now fortunately corrected) in both the United States and Canada have harmed plant economics. Lack of standardisation of reactor designs was a major weakness in the United States, which in turn was caused partly by the fragmentation of ownership – in contrast to France where one utility owns and operates all the nuclear plants. The failure to resolve the waste issue has also been a negative feature, with continued delays to the Yucca Mountain repository. Nevertheless, 103 reactors are now operating very well in the United States, with a further one expected to come back on line in 2007. Their generation costs are lower than the competing power technologies and most, if not all, are expected to receive license extensions up to 60 years.

Asia has gradually become the most important area for nuclear power, particularly looking ahead. Japan has embraced nuclear power for similar energy security reasons to France, but without pushing the nuclear share up so far. Recent reactors have been built on schedule and to budget, although public acceptance has been damaged by a number of comparatively minor incidents on safety and its verification. It can be argued that Korea has perhaps the best-balanced and planned power generating sector in the world, with nuclear now taking around 40% of the generating mix, combined with coal, gas and oil. From a base of imported technology, Korea can now design and build its own reactors and is on the verge of exporting these to other countries. China and India are now the obvious growth markets for new nuclear reactors and both have announced ambitious plans. China has used a mix of indigenous and imported technology whereas India, barred (up to now) from international nuclear commerce through its weapons programme, has had to rely mainly on domestic technology.

International co-operation

The world has now reached the point where a further expansion of nuclear power depends crucially on enhanced international co-operation. President Eisenhower's 'Atoms For Peace' was the foundation of the civil industry over the past 50 years and led to the initial spurt of nuclear power with the International Atomic Energy Agency (IAEA) policing the necessary safeguards against civil materials being put to military uses. We now need to move beyond this, as several weaknesses in the existing regime are quite obvious.

The non-proliferation arrangements have been sound to the extent that few new countries, beyond the original five, have acquired nuclear weapons while the ability of terrorist groups to misuse nuclear technology remains unproven. Yet concerns about proliferation are today as strong as ever, reinforced by the recent troubles over Iran and North Korea. A way has also to be found to bring India and Pakistan within the existing non-proliferation regime or, at the very least, to allow them to co-operate more closely with international partners rather than be isolated on the outside. Another clear weakness is with the management of used nuclear fuel. Under the existing regime, this is a national matter with each country required to find solutions for its own waste materials – which will eventually logically involve a national repository in each country. This is economically absurd and will therefore serve to prevent a resolution to probably the biggest issue with which the industry has difficulties over public acceptance. Finally, although the number of trade restrictions in the nuclear fuel cycle has diminished over time, there still remain areas where lack of bilateral nuclear co-operation agreements or national protectionism prevents a better flow of materials and technology.

There are already complementary initiatives in place to address these issues, led by the IAEA, the US and Russia. The director general of IAEA, Dr El Baredei, has proposed establishing regional centres for uranium enrichment, used fuel reprocessing and final disposal of wastes, all closely monitored under IAEA safeguards. These would service the world nuclear industry with fuel throughout the full cycle, particularly for those countries new to nuclear power. This should meet non-proliferation objectives but also avoid the excessive nationalism of the previous regime, which was both non-economic and led to countries avoiding the difficult decisions on waste. The Global Nuclear Energy Partnership (GNEP) proposed by the United States fits quite comfortably with this (DoE, 2006). Although other countries may doubt whether the United States has any right to take on leadership in these matters, the ideas are constructive. A lot of attention has been placed on the apparent conversion to the reprocessing of used fuel within GNEP (perhaps as a reaction to the difficulties being experienced in getting Yucca Mountain into operation) but the degree of international co-operation proposed at all levels of the fuel cycle is very welcome. It makes good sense to have enrichment, reprocessing and waste disposal facilities located at only a very limited number of locations throughout the world, for both non-proliferation and economic reasons (the economies of scale in all areas of the nuclear fuel cycle are considerable). The Russians have also proposed developing sites for major regional facilities for these areas within their own territory, possibly starting with waste repositories. Moving used nuclear fuel is currently almost impossible owing to historic restrictions, but it is conceivable that a market may eventually develop in this, allowing disposal in countries other than where it has been created.

Two other initiatives worth mentioning are the recent (2006) US–India agreement on nuclear commerce and also mounting co-operation on the next generation of reactors. The former provides a way of bringing India back into the nuclear fold, at a time when it has a great need and evident desire to develop a huge nuclear power programme. It must achieve ratification by Congress and also the consent of the other parties to the Nuclear Suppliers Group, but it at least provides a constructive route to solving an obvious problem. Indeed, the whole non-proliferation regime must find ways of addressing the other outstanding issues of Pakistan, Israel as well as Iran and North Korea. Co-operation on developing the next generation of reactors exists within two increasingly complementary international programmes, that of Gen IV (led by the US) and the INPRO project within the IAEA (led by Russia). These are considering a range of advanced reactors which will have much-enhanced proliferation-resistance but will also operate much more economically – in particular using much less fuel and generating less waste. Indeed, they may largely be fuelled by reprocessing the used fuel from current reactors and using the separated plutonium and uranium. They will also address the perceived need for nuclear reactors for seawater desalination and hydrogen production. These reactors will come into operation in the period after 2020 – up until then, it is reasonable to assume that new reactors will be evolutionary versions of current reactor types, with similar characteristics.

Worldwide nuclear revival

Although nuclear technology has now arguably reached a high level of maturity (the main reactor design in use throughout the world goes back to the 1950s) the industry itself has clearly not. Because of some technical failures, falling fossil fuel prices and the failure to explain itself well to the general public, its rapid expansion got cut off in the 1980s. There is currently a significant increase in the attention being given to the possibility of new nuclear build in many countries – in some cases in those currently without reactors, such as Poland, Indonesia and Vietnam. The reasons for this are partly economic, but perhaps more connected in most countries with security of supply and nuclear's possible role in greenhouse gas abatement. But talk is cheap and there remains a significant risk that new build programmes will amount to little more than this.

This is unlikely due to the arguments of those opposed to nuclear power. These are increasingly being shown to be weak and without substantial foundation. It may be premature to say that the intellectual battle in favour of nuclear power has been won, but it is increasingly looking that way.

Provisions which have been put in place to guarantee plant security from terrorists and operational safety are widely seen as very effective, while the international non-proliferation regime is developing but has already been effective in practice. Although the latest plant designs have yet to prove beyond all doubt that they can be highly economic, the economic success of current operating reactors suggest that this is highly likely. Cheap and reliable power, without likely price spikes, over many years is a key advantage of nuclear power plants and the message is slowly getting across that this is so. What is needed is the courage to get over the initial period of pain of high initial capital costs to enter the 'land of milk and honey' in subsequent years, where nuclear plants can be almost 'money machines' for their owners. The waste issue is perhaps still the most difficult one, but combinations of national political will to push solutions and the international initiatives mentioned above should go a long way towards solving it. The uranium resource issue which is often brought up nuclear opponents is a complete red herring, as any understanding of the reserves figures, the significant increase in exploration now occurring and the likely changes in reactor technology beyond 2020 will quickly show. The argument that resources will eventually get stretched and involve exploiting high carbon-emitting uranium mines is again nonsense – the average grade of deposits being exploited is actually rising and even if low grade reserves are to be developed, techniques such as in situ leaching (ISL) are fundamentally low carbon-emitting. Finally, decommissioning costs are not material in new nuclear build decisions (as are not the major waste management costs, for similar reasons). The costs will be incurred so far in the future that the plant owner will not have to put away huge sums each year to fund them – provided they start early on, the fund will eventually accumulate nicely (as a pension fund indeed should).

The real challenge to a nuclear revival will come from either a lack of political will or a lack of guts from those charged with making decisions on new generating capacity. We can see each of these by referring to the United Kingdom and the United States, in both of which a new nuclear build revival is crucial as a lead to the remainder of Europe (UK) and the rest of the world (US).

In the United Kingdom, the government has to demonstrate the political will to support new build. It must develop the regulatory system to ensure that reactors can be approved and built in a reasonable time, finalise policies on waste management and decommissioning (so the plant owners know exactly the extent of the financial provisions they must make) and satisfy the requirements on plant security and nuclear liability. The United States has got rather further ahead in most of these, so the issue there is more 'who is ready to invest?' Nuclear power plants are highly complex projects and it is almost understandable if a power utility chief executive holds up his hands in horror at the prospect and says, 'let's build a gas or coal plant instead'. This may be a much safer business decision, so some courage is needed, at least for the first investor to commit. It may need some initial government subsidies to encourage them, such as the loan guarantees and production credits already proposed, but these should be sufficient to get the ball rolling. After that, the nuclear sector should be able to stand on its own feet and show that it can indeed generate a huge quantity of power economically and environmentally-soundly, while contributing to national and regional energy security of supply.

References and further reading

DoE (2006) Global Nuclear Energy Partnership website, US Department of Energy: *http://www.gnep.energy.gov.*
World Nuclear Association *website http://www.world-nuclear.org.*

Part VI Conclusions

15
Can Nuclear Power Ever be Green?

Jonathan Scurlock

This chapter reflects upon the history of the environmental movement as well as the UK energy policy context and global resource base, to pose a number of questions. Can nuclear power be considered a 'green' technology if it is part of a wider response to the threat of climate change? Should the green movement engage in a reappraisal of its core values in order to address this issue? Is an anti-nuclear stance fundamental to green ideology, or can it be discarded as 'excess baggage'? Is it possible for environmentalist opinion to converge with that of the nuclear industry, and can they be reconciled as 'separate but different', possibly even complementary?

The policy context

Nuclear power has been 'off the menu' in Britain since the privatisation of the electricity supply industry in the late 1980s, at the height of the Thatcherist approach to public investment. The UK government's last review of nuclear power was published in the mid-1990s (DTI, 1995), and the Energy White Paper of February 2003 stated that 'current economics ... make it an unattractive option' (DTI, 2003). For many greens, the 2003 White Paper represented a visionary break with past policies, but with hindsight, its goals were compromised by a failure to set targets for 2020, as well as the absence of effective follow-up policy measures, both regionally and nationally. This, together with the limited progress made by renewable energy and energy efficiency towards mid-term cuts in carbon emissions, left the door open for the UK nuclear power industry to demand a further energy review, to examine its own role in a low-carbon economy from 2020 onwards.

There also appears to have been a struggle for dominance of UK energy and climate change policy, between ministers and civil servants in two government departments notorious for their lack of co-operation, DEFRA (the Department for Environment, Food and Rural Affairs) and DTI (the Department of Trade and Industry). Neither has been particularly enthusiastic about the costs, cumulative risks and public perception of nuclear power, but there is clearly a realisation that energy efficiency and renewables in Britain are failing to deliver low-carbon energy services quickly enough.

At the same time, frustration with the snail's pace of international action on climate change has led some environmentalists to toy with the idea of tolerating nuclear as a temporary (or even permanent) fix to avert a much worse threat to the planet. Other greens are horrified at such apostasy. Since the late 1960s, nuclear power has represented to them the worst aspects of the military-industrial complex, Big Government and Big Industry (Herring, this volume, Chapter 3). Many still feel that opposition to the 'atomic menace' is one of the building blocks of the environmental movement.

A pause for thought

The 2006 review of the role of nuclear power in UK energy policy was widely anticipated, but it seems to have provoked a re-appraisal of the very notion of 'green' values. Is it possible to dump some of the ideological baggage overboard? At a time when some environmentalists appear more concerned about local landscapes than the Earth as a whole, perhaps the environmental movement should engage in its own internal review – posed by the question 'can nuclear power be (or ever become) a green technology?'

As explained in Chapter 3, the modern green movement grew largely out of the counter-culture revolution of the late 1960s in the United States and Western Europe. This was itself rooted in the disaffection with post-World War II values felt by demobilised servicemen and young intellectuals in the 1950s. Other, older green threads lead back to more right-wing, ruralist conservative movements such as the Distributism of the 1930s and Malthusian philosophies that also saw a revival in the 1970s. Green politics worldwide in the 21st century has mostly distanced itself from such balkanist/apartheid viewpoints, and these days embraces the so-called four pillars of ecological sustainability, global social justice, local grassroots democracy and non-violence. The turning point for many was the UN Conference on the Human Environment, held in Stockholm in June 1972, which confirmed an international dimension to environmental awareness.

How much of this recent history is still relevant to the current threats to Planet Earth? In our rapidly changing world, climate change has grabbed most of the headlines, but we are also preoccupied with the impact of terrorism on the security of our Western lifestyle (and indirectly, the effect of counter-measures on many of the conveniences we take for granted). Meanwhile, the global justice lobby, itself an arm of the green movement these days, offers a more humanist perspective on how to combat the roots of terrorism. And the concerns of the 1970s generation about long-term resource depletion and sustainability have not gone away, either.

As pointed out more than 30 years ago (Meadows et al., 1972), it may not be possible to sustain worldwide economic growth for more than a few additional human generations without exceeding our planet's 'carrying capacity'. Although we may be able to devise technological methods to increase the efficiency of resource use, there is little doubt that the Earth does ultimately have a finite 'safe working load'. Humanity has no long-term future on this planet unless we learn to temper some of our material expectations and remain well within these natural constraints. Aside from our recent preoccupation with energy technologies and the global carbon cycle, perturbation of the global nitrogen cycle and local over-exploitation of freshwater reserves are already giving cause for concern. Sooner or later, the human race will have to shift beyond the realm of economic globalisation and into an era of wholesale planetary management.

Given the revival of doom-laden views about the future, perhaps it is not so strange to find a pro-nuclear splinter group on the green fringe, such as EFN (Environmentalists for Nuclear Energy), an association founded by the French physicist Bruno Comby and supported by the likes of James Lovelock. In the 1950s and early 1960s, at the dawn of the 'atomic age', there was support for nuclear power from at least some of those campaigning against nuclear weapons, who approved of the 'swords into ploughshares' rhetoric. However, there is a startling contrast between the 1953 'Atoms for Peace' speech by President Eisenhower, who called for the 'universal, efficient, and economic usage' of nuclear power, and the 21st century demands by the White House and its allies to limit the international spread of uranium enrichment technology. At the same time as trying to launch an atomic comeback in OECD countries, we are trying to stuff the nuclear genie back into the bottle in the Middle East. Where is the justice in that?

Technology assessment

It is inevitable that technology shapes society to some extent – look at the impact of motorised transport, or more recently the Internet and mobile phones. But it is also possible for society to deliberately shape technology – for example, weapons control treaties, the Montreal Protocol on ozone depletion, and (of course) mostly recently the Kyoto process. Such 'technology assessment', or perhaps more precisely 'active techno-logical choice', provides a counterpoint to the purely technocratic point of view, by recognising that technological progress does not come without ethical implications.

Within the domain of energy technologies, it may be argued that human-ity is challenged to match the rhythms of human energy and activity with the rhythms of the Earth's own energy flows, in order to attain a long-term sustainable economic system for our planet. Utilisation of some energy resources may compromise the ability of future human generations to sus-tain their own needs, while others may appear to be relatively benign. Although we cannot fully predict the desires of future generations, we do have historical models that enable us to comprehend certain ethical issues such as resource depletion and pollution. The conventional definition of 'sustainable energy' is one that is not substantially depleted by continued use, does not entail significant pollution or other environmental problems, and does not bring about health hazards or social injustices (Alexander and Boyle, 2004).

However, we should take care that the process of technology assessment does not fall hostage to naked conservatism, given that change is an inevitable process. Thus, to overemphasise the impact of non-depleting renewable energy technologies such as wind power on the 'sacred land-scapes' of Britain is to ignore the progressive land-use change which has been a feature of British history since at least the 16th century. Land enclosure, modern agriculture and the building of wooden ships have all played their part in changing the landscapes of the industrialised world – and yes, each successive generation has resisted change, only to roman-ticise the new landscapes that emerged from it. Perhaps it was a historical inevitability that we, the British, had to cut down most of our ancient forests in order to become a successful military and trading nation. Perhaps we also have to accept a certain environmental trade-off in measuring up renewable energy against nuclear power, if we are to show the rest of the world the path towards sustainability.

Nuclear's green credentials?

To fully assess whether nuclear can be considered 'green', it may be neces-sary to do more than simply estimate and compare full life cycle emissions of carbon dioxide between energy technologies. Apart from carbon emis-sions, there are other aspects of sustainability to consider – but that is not to underestimate the value of a truly independent carbon life cycle analysis, accepted by nuclear proponents and the environmental movement alike, and based upon both present technology and resources and those projected for, say, the year 2050.

Most authorities agree that conventional resources of uranium (comparatively high-grade ores) are enough to last only 75–100 years based on current technology and current installed capacity. Higher-cost uranium, derived from lower-grade ores, would not significantly impact overall generating costs – but the carbon emissions from the 'front end' of the fuel cycle would increase. Even allowing for (very gradual) technological improvements in reactor fuel economy, blending of depleted uranium with weapons-derived highly-enriched uranium, and a revival in reprocessing of spent fuel, fission reactors face an increasingly costly and carbon-intensive fuel cycle, if nuclear's share of world electricity generation is to be even maintained at its present level. Of course, if and when more reactors came on line, then these could supply some of the power for mining and fuel fabrication, but the energy requirement and carbon emissions would continue to grow, while the fuel resource would go on diminishing.

As mentioned in Chapter 7, the worst-case scenarios presently modelled, based upon the use of low-grade ores, suggest that overall carbon emissions from nuclear power in the future could actually be greater than for current gas-fired generation, and would continue to increase over time. This may seriously erode the present justification for OECD governments to create the financial environment for a thriving carbon-saving nuclear industry, even if we could defend the refusal to transfer the technology to 'unsuitable' countries in the interests of non-proliferation.

Leaving aside the ecological sustainability of storing even modest quantities of long-lived radioactive wastes, future energy technologies, whether based on nuclear or renewables, must also address the 'Limits to Growth' conundrum (Meadows et al., 1972). It still puzzles me to read of respected nuclear scientists and technocrats, past and present, expressing bewilderment at the environmentalist perspective. I am sure they honestly believe (or believed) that every resource constraint, every waste management problem, to the year 2100 and beyond, may be solved by injecting ever-larger amounts of electrical power into the global economy – but they are painting a scary picture of the future (Kahn et al., 1976).

It has been argued that a system using fast breeder reactors could, in theory, produce energy for millions or even billions of years from extremely low-grade unconventional sources of uranium, such as sea water (Cohen, 1983). But breeder reactors have proved complex and expensive so far, and their fuel cycle represents a further turn of the proliferation spiral.

Furthermore, the fuel-cycle technology and environmental consequences of processing such colossal quantities of sea water to extract uranium would make contemporary pollution problems with radioactive mining tailings look like a picnic.

And future energy technologies also need to address the other 'pillars' of the green movement. An energy technology whose resources and intellectual property rights are unequally distributed around the world (like oil, coal to a lesser extent, and most certainly nuclear) is not going to advance social justice or contribute much to local empowerment. If it requires a draconian security system to protect its capital assets (or the public from health hazards) then it is hardly compatible with the principle of non-violence, either.

Renewable resources – is nuclear needed?

Another way to consider nuclear power's green credentials is to explore whether any kind of high-energy technology is indeed a necessary part of our future. How far could humanity push the boundaries of the Earth's energy services without resorting to nuclear? Some brave attempts at such forecasts have been made (Goldemberg et al., 1988) to demonstrate how renewables and energy efficiency could bring the standard of living of the world's poor up to that of, say, 1970s Sweden without compromising continued economic growth for the rich. Undoubtedly, there are finite limits to the economically and environmentally exploitable resources of renewable energy – but the range of technologies is diverse, and many have considerable further potential for development. What is most likely is that the real (income-adjusted) cost of energy itself will have to increase over time, in order to provide an economic driver for the necessary improvements in efficiency of end-use. However, the real cost of the services derived from energy would probably remain constant for the rich world, and may actually decrease for the world's poor, as their incomes rose and more energy-efficient devices 'trickled down' (Goldemberg et al., 1988). A nuclear utopia of cheap energy and profligate consumption now seems an unlikely scenario, at least on Earth.

Solar radiation striking the Earth on an annual basis is still equivalent to over 10,000 times global energy consumption. The amount of 'natural' energy flowing through particular pathways is somewhat less, but still substantial. For example, the worldwide wind power resource (much of it offshore, confined to the shallower regions of the continental shelves) is equivalent to several times present world electricity demand, and a UK Marine Foresight Panel recently estimated that harnessing just 0.1% of energy in the oceans would meet world energy use five times over. Likewise, annual photosynthetic energy storage is around ten times present world energy use (although the accessible biomass energy resource is somewhat less, given the ecological value of much of this resource, as well as other competing uses for food, fodder and fibre).

Biomass energy provides a classic example of a 'green' energy resource. Finite, but renewable, its full potential has yet to be realised because of poor market development in many parts of the world, despite its comparative simplicity and accessibility. Supply and utilisation of modern biomass fuels may bring real benefits, through sustainability, local incomes and empowerment, as long as development of markets is accompanied by some kind of checks and balances. Damage to valued ecosystems through expansion of cash crops has indeed been continuing since the 1960s, but largely as a result of the growth of a poorly regulated and unjust world food industry. Adding more fuel crops to the present mix of fodder and food crops, as an explicit instrument of environmental policy, provides an opportunity to improve the situation. Likewise, domestic demand for bioenergy crops in wealthy countries like the United Kingdom may allow the partial recovery of some landscapes from intensive food cropping. Furthermore, the notion that many poorer countries may now be able to produce something that the rich world really wants (i.e. low-carbon fuels) is seen by some advocates as a chance to improve the balance and terms of trade.

A time to build bridges – or burn them?

Can nuclear and renewables be reconciled as 'separate but different', possibly even complementary, responses to global warming? After all, nuclear power and energy-capturing renewables such as wind turbines share many features in common – relatively high capital costs and comparatively low operating and maintenance costs, in addition to their low carbon emissions compared with fossil fuel-fired power generation. It is ironic, too, that many of the shortcomings of government energy policy that have led to a withering of the nuclear vine have also held back the development of renewables.

How will they measure up as the 21st century unfolds? It will be an exciting race to watch. If nuclear power continues to stagnate, current projections suggest that worldwide installed wind power capacity will overtake nuclear *capacity* some time during the next 10 years, although since nuclear has a higher load factor than wind, nuclear electricity *production* will not be surpassed by wind until around 2020. In the EU

alone, wind capacity is presently about one-quarter of nuclear, and electricity generation about one-eighth. In three countries with significant
nuclear industries (Spain, Germany and India), wind generating capacity
is already on a par with nuclear – a technology that has been around for
decades longer. It is harder to compare the degree of employment, direct
and indirect, between the two industries, but it is also plausible that wind
employment worldwide may be the greater by about 2010.

As we observe this technological contest, is it possible for environmentalist opinion (apart from that of factions such as EFN) to converge
with that of the nuclear industry? For example, Peter Harper and Paul
Allen of the Centre for Alternative Technology, Machynlleth (a hub of the
UK green movement from the 1970s onwards), while still resolutely anti-
nuclear, have explored the idea of supporting the use of nuclear fuel
blended from depleted and highly enriched uranium, derived from
nuclear weapons decommissioning, to buy more time for deployment of
renewables. 'The worst possible nuclear disasters are not as bad as the worst
possible climate change disasters', they argued, in a commentary on a
conference hosted by EFN member James Lovelock (Harper and Allen,
2004). Subsequently both authors rejected the idea of keeping nuclear as
an interim strategy, and reconfirmed their continued opposition to the
nuclear option. Most of the rest of the UK green movement has not even
wavered.

Some of the green resistance to nuclear power might however soften
if the industry could deal with four persistent issues – but they all entail
likely higher costs and a diminishing of its low-carbon potential. Truly
independent regulation of nuclear facilities by stricter authorities would
be the first; and a more secure and completely reversible system for storing spent fuel and/or wastes is second in line. To counter proliferation,
there should be a worldwide ban on fuel reprocessing, and uranium
mining should also be subject to much greater environmental scrutiny.
Whether the nuclear power industry, in the United Kingdom or worldwide, would submit to such conditions in the interests of 'environmental
détente' is an interesting question.

Yet other environmentalists claim that they are no longer opposed 'in
principle' to nuclear power, and that the arguments against reviving the
nuclear option are no longer ideological, but technical and financial.
Although operating costs have been reduced, the nuclear industry is still
crippled by high capital and finance costs – and most of all by its inflexibility. Its heyday in the 1960s and 1970s was at a time of 'predict and
build', and it struggles to find a place in a world of private-sector finance
and project management. However, the private sector is essentially a

technology-neutral agent that does not indulge in activities such as technology assessment. If the market conditions are right, it will invest. Thus nuclear advocates point to the special-purpose vehicle set up to finance the Olkiluoto plant in Finland, suggesting that major British energy users may support a similar consortium in the United Kingdom if the government creates the right risk and regulatory conditions for new nuclear build. Ironically, some of the Finnish intensive energy users are the forest product and paper industries – themselves a source of significant investment in (renewable) biomass energy.

But is the nuclear emperor really wearing fine new clothes today – or are these projected low costs coupled to low carbon emissions just an illusion? Certainly, the costs of new nuclear build in the United Kingdom are highly speculative, given the lack of recent construction experience. The last British reactor, Sizewell B, was finally ordered in 1987 (nearly 19 years ago), and completed in the mid-1990s at a cost of more than twice its original budget. Nuclear advocates have lately been quoting a report on generating costs by the Royal Academy of Engineering (RAE, 2004), but this bases its nuclear costs on optimistic projections only (essentially the same numbers submitted by British Energy and BNFL to the 2001 Energy Review by the UK government's Performance and Innovation Unit). Both of these sources contain similar proposals – to construct a new series of 8 or 10 advanced light-water reactors based on American or European designs. At least a construction programme on this scale has some chance of meeting its own expectations, but a more likely outcome would be a poor 'British compromise', whereby one or two hugely expensive units would be built on existing nuclear sites such as Hartlepool, Hinkley Point or Sellafield, with all the waste and diversion of resources that entails.

Moreover, even a doubling of current British nuclear capacity from 12 to 24 GW has been estimated to reduce national greenhouse gas emissions by barely 8%, since electricity constitutes only about one-third of total energy use (FoE, 2004). The renewable energy industries have already set more challenging targets for low-carbon energy by the year 2025, more than enough to plug the gap left by declining nuclear generation (BWEA et al., 2005). If renewable energy's problems with private sector finance and planning are delaying it from making a difference to the climate, new nuclear capacity will face even more difficulties.

Most environmentalists argue that a diversity of small and medium-scale renewable energy projects (wind, biomass, solar, others) offer the best prospect of building a low-carbon energy economy rapidly and flexibly. I agree: it appears already very unlikely that, at least in the United Kingdom, we could build and commission nuclear power stations fast enough to replace those soon to be retired (SDC, 2006). Calls by the World Nuclear Association and others for a doubling of current nuclear capacity (in Britain, the United States or worldwide) are a shot in the dark, at best.

Perhaps the 'least worst' nuclear option for the United Kingdom, and arguably worth the support of the environmental movement, is for there to be no new nuclear build, but for the lives of the most suitable advanced gas-cooled reactors (AGRs) to be extended. Indeed, British Energy announced in late 2005 its proposal to operate the Dungeness 'B' AGR for an extra 10 years beyond 2008. Together with the Sizewell 'B' pressurised water reactor, a limited fleet of revitalised AGRs would maintain a few gigawatts of nuclear capacity beyond 2030, allowing longer for energy efficiency and new renewables to be deployed. However, many of these graphite-moderated reactors are probably unsuitable for re-licensing, due to ageing and cumulative radiation damage to the moderator material.

Conclusions

Without significant compromise on the part of the nuclear industry world-wide, most environmentalists will reject the notion that nuclear can ever be considered a 'green' technology. Time and again, nuclear power has proven to be too slow, too expensive and too limited in its scope to be an effective response to global warming. Whether it can, and whether it will nevertheless be pushed ahead, remains to be seen.

After all, renewable energy has made some progress despite the lack of significant support. I firmly believe that, given a chance, renewables could fulfil their 21st-century promise. Although nuclear power may play a part in meeting the climate challenge, it will be a modest and diminishing role. Perhaps the main benefit of the present-day arguments over the role of nuclear in a low-carbon future will be to bring old arguments back into sharp focus.

Nuclear power may yet find a specialised 'sustainable' role for itself. Proponents of the manned exploration and eventual exploitation of space have long argued that small nuclear fission reactors are the only way to meet immediate needs for electrical power and heat, e.g. for sustained human exploration of Mars (Zubrin, 1996). Small isolated applications where no other technology is available (such as present-day nuclear submarines, as well as space exploration where solar energy flux is limiting) are arguably an efficient and 'sustainable' use of resources, assuming that the resulting modest amounts of long-lived nuclear waste can be safely stored for eventual transmutation. Transmutation of long-lived or highly radioactive isotopes (by bombardment with neutrons or other high-energy particles) may also eventually prove to be the saving grace of nuclear fission technology – but only if the total amount and cost of safe storage, transport and waste processing is kept to a minimum. In an extraordinary overturning of the arguments put forward by proponents of both nuclear and large-scale fossil-fuel power generation – that renewables would only ever play a marginal role in niche applications – it appears that the future of the nuclear genie may be confined to various niche roles, where its high energy density and power density can justify the expense and risk of radiation shielding and managing nuclear waste.

Over the past century, the promise of nuclear science and technology has absorbed the enthusiasm and skills of many people who felt that they might be benefiting the future of the human race. Not all of this expertise will be lost to the world: much of it can be redirected to the new emerging technologies. Indeed, if renewables and the other technologies of 'sustainability' are to expand on the scale that now seems to be necessary, we will need all the expertise we can find.

References

Alexander, G. and G. Boyle (2004) Introducing renewable energy. In: *Renewable Energy: power for a sustainable future*, 2nd edn. (G. Boyle, ed.). Oxford University Press, Oxford.

BWEA, REA and others (2005) The Renewable Energy Manifesto: an action plan to deliver 25% of the UK's energy from renewables by 2025. Press release by eleven UK renewable energy trade associations, 21 February 2005. British Wind Energy Association/Renewable Energy Association, London.

Cohen, B.L. (1983) Breeder reactors: a renewable energy source. *Am. J. Phys.* **51**, 75–76.

DTI (2003) Our Energy Future – Towards a Low Carbon Economy, Energy White Paper, Cm 5761. Department of Trade and Industry, London/ HMSO, Norwich.

DTI (1995) The Prospects for Nuclear Power: Conclusions of the Government's Nuclear Review, Cm 2860. Department of Trade and Industry, London/HMSO, Norwich.

FoE (2004) Why nuclear power is not an achievable and safe answer to climate change. Briefing pamphlet, Friends of the Earth, London, www.foe.org.uk.

Goldemberg, J., T. Johansson, A.K.N. Reddy and R. Williams (1988) *Energy for a Sustainable World*. Wiley Eastern, New York/Delhi.

Harper, P. and P. Allen (2004) New Ethical Perspectives in Energy Policy. Commentary on the Conference on Gaia and Global Change, Dartington Hall, Devon, UK, June 2004, extract quoted in *Renew* 152 Nov–Dec 2004, p. 33.

Kahn, H., W. Brown and L. Martel (1976) *The Next Two Hundred Years: a scenario for America and the world*. William Morrow and Co., New York. 241 pp.

Meadows, D.H., D.L. Meadows, J. Randers, and W.W. Behrens III (1972) *The Limits to Growth*. Universe Books, New York.

RAE (2004) *The Costs of Generating Electricity*. Royal Academy of Engineering, London. 56 pp.

SDC (2006) The Role of Nuclear Power in a Low Carbon Economy. Sustainable Development Commission, London. 24 pp.

Zubrin, R. (1996) *The Case for Mars*. Simon & Schuster, London, 328 pp.

16

Nuclear Renaissance Requires Nuclear Enlightenment

William J. Nuttall

Introduction

Despite this author having considered previously the prospects for a *Nuclear Renaissance* in Europe and North America (Nuttall, 2005a), it is important from the start of this chapter to concur with others in this volume and to concede that nuclear power cannot yet be regarded as *sustainable* in a formal sense. Rothwell and van der Zwann have examined the sustainability of current light water reactor (LWR) systems in some detail and they conclude that while LWR systems are consistent with the intermediate form of sustainability over the foreseeable future when one considers environmental externalities and social externalities associated with health and safety, LWRs fail in respect of non-renewable resource depletion, a lack of effective institutions to restrict proliferation and the capital-intensive economics of new build (Rothwell and van der Zwaan, 2003). The failings identified by Rothwell and van der Zwann might be overcome in time as new nuclear reactor technologies are deployed, novel (e.g. thorium-based) fuel cycles are developed and financial and regulatory structures improve. Rothwell and van der Zwann, neglect however to consider one of the greatest challenges to the social sustainability of nuclear power – social acceptance.

Arguably all large-scale energy sources currently fail to achieve sustainability in some way or another. In respect of poor sustainability nuclear power is no exception, but in numerous other ways it is. Recognising that nuclear power is special Gordon MacKerron has suggested that it must become 'ordinary' if it is to find an enduring role in western electricity systems (MacKerron, 2004). One important aspect of the lack of ordinariness in nuclear power is unalterable – its historical association with the development of nuclear weapons and the Cold War. The synergies between nuclear weapons development, naval propulsion systems and commercial nuclear power are powerful and undeniable. In fact it is the synergy between naval propulsion and the successful emergence of light water reactors that is most important in the history of nuclear electricity. In many ways this history is paralleled by the synergy between the development of gas turbine technology for electricity generation and military aerospace research and the development of jet into jet engines. As for the link between nuclear weapons and nuclear power, for the countries with permanent seats on the United Nations Security Council nuclear weapons development predated the development of nuclear energy systems. It is arguable that some later members of the nuclear club, such as India and Pakistan developed nuclear weapons programmes in concert with their civil nuclear energy projects. These states used a nuclear energy infrastructure and knowledge base to assist with the separation of plutonium and the enrichment of uranium to provide materials for fission weapons. Both states remain outside the international non-proliferation regime. It is incorrect, to regard nuclear weapons developments as an inevitable consequence of nuclear energy programmes. For instance, neither of the key sensitive nuclear materials highly enriched uranium (HEU) nor separated plutonium are required for the operation of a commercial nuclear power programme. Global moves towards a nuclear renaissance, such as might be required to militate against the global threat of anthropogenic climate change would appear to require increased internationalism and globalisation. Michael May and Tom Isaacs have argued forcefully in such terms for a strengthening of global non-proliferation measures (May and Isaacs, 2004). While the bottom up emergence of local initiatives is a possible route to sustainability, indeed it is the dominant paradigm for renewables, the proliferation risks of nuclear power imply that the nuclear approach to a low carbon future must be via a large-scale internationalist approach if proliferation and terrorism risks are to be minimised. This leads us to recognise that public attitudes to centralised authority versus decentralised decision-making are central to the future of nuclear power. Malcolm Grimston has touched upon these issues when he argues that a key difficulty of nuclear power is that it is poorly matched to modern preferences for local, or even individual, control (Nuttall, 2005a, p. 78). In extremis such a thesis posits that it is not cost, safety or environmental performance that is key to public attitudes to energy options, but rather the nature of the individual's control of technology. A micro turbine or Stirling engine in one's kitchen fits the *Zeitgeist* better than a nuclear power station in the next county. These possible aspects of public acceptance need to be tested carefully in future public attitudes work.

Does nuclear fission lead to technocracy?

The relationship between nuclear power and public attitudes prompts the more general question posed by Langdon Winner: 'do artefacts have politics?' and the particularly challenging and stronger question 'do technologies shape or determine political action?' (Winner, 1986). In his book *The Whale and the Reactor*, Winner challenges the prevailing orthodoxy that holds that it is absurd to attribute political power to technologies assembled by man from raw and inanimate materials. Indeed this prevailing attitude implies a world view that technology is socially constructed rather than that society itself is technologically constructed. Winner argues that not only do artefacts have political consequences but that certain technologies do indeed imply forms of social and political organisation. Winner gets to the nub of our concerns when he quotes Jerry Mander:

> if you accept nuclear power plants, you also accept a techno-scientific industrial-military elite. Without these people in charge, you could not have nuclear power. (Winner, 1978)

This chapter has tended to the conclusion that technologies can prompt a need for new political and social decisions but contest the view that the outcomes of such deliberations are in any way inevitable or predetermined. Therefore one might ask: is it possible to have a nuclear power industry that is not governed by Jerry Mander's techno-scientific industrial-military elites? Might a nuclear power system be constructed that exists only at the pleasure of the people and which is shaped by their concerns?

Winner posits that because uranium is a finite resource commercial nuclear power will inevitably move to a plutonium economy. Over the long term proliferation will be inevitable and to militate against such risks society must move to an Orwellian surveillance state. These concerns and the 'Atomic Priesthood' concept developed by Thomas Sebeok[1] imply a surveillance society separating a technocratic nuclear elite from an ordinary population living in ignorance of such matters. Winner argues that attempts to boost public acceptance of nuclear power cannot yield protection against the drift to the plutonium surveillance state. He argues:

> Yes, we may be able to manage some of the 'risks' to public health and safety that nuclear power brings. But as society adapts to the more dangerous and apparently indelible features of nuclear power, what will be the long-term toll in human freedom? (Winner, 1986)

Whether the presence of separated fissile materials will yield the totalitarianism feared by Winner or simply require stronger international oversight as proposed by May and Isaacs is partly a matter of individual political perception. What is clear is that the notion that nuclear power risks eroding democracy, privacy and individual liberty is well established. A particularly pessimistic vision forms the basis of Robert Jungk's book *The Nuclear State* (Jungk, 1979). He asserts that nuclear power represents a fundamental tipping point in the evolution of human society. He warned in 1979:

> The totalitarian technocratic future has already begun. Chances of preventing it still exist, but time is short. A peculiarity of atomic development stems from the fact that it can be arrested only up to a point of no-return. Once that point is reached it is impossible to stop. This 'irreversibility' is an entirely new phenomenon in history… When the number of installations and waste disposal units has passed a certain stage, the necessity for strict surveillance and control will leave their mark permanently on the political climate. (Jungk, 1979, p. xiii)

Robert Jungk was a prominent futurist and opponent of authoritarianism. It is interesting to note the special attention that he gave to nuclear matters during his career. He died in 1993 and so now is unable to advise us as to whether society has indeed reached its *point of no return*.

The warnings of Winner, Jungk and others are important at a substantive level as they refer to the future of our liberal societies. It is not the purpose of this chapter to seek to assess whether they will be proved right, rather these issues are raised as they form an important part of legitimate public concern regarding nuclear power. Several prominent thinkers have argued that nuclear power erodes freedom, however the converse view is also worthy of consideration. Perhaps nuclear power may even have a positive role in preserving liberal society. If the thoughtful public is concerned that energy and environment policy has the potential to alter society, then perhaps there is a benefit in the public being encouraged to ask where the greatest threats to liberalism really lie. The threat of the plutonium society has now been well articulated for several decades. In recent years the public has learned to consider the impacts on our society that will arise from anthropogenic climate change. This may have profound importance for the public acceptance of nuclear power.

Nuclear power is an almost zero greenhouse gas electricity source contributing roughly 16 per cent of global electricity (Hore-Lacey, 2003). The

UK Royal Commission on Environmental Pollution stressed the importance of carbon dioxide emissions reduction when in 2000 it noted:

> For the UK, an international agreement along these lines which prevented carbon dioxide concentrations in the atmosphere from exceeding 550 ppmv and achieved convergence by 2050 could imply a reduction of 60 per cent from current annual carbon dioxide emissions by 2050 and perhaps of 80 per cent by 2100. These are massive changes. But the government should implement short, medium and long term strategies which are sufficiently coherent and effective to achieve these reductions. (RCEP, 2000)

Any measures to achieve 60 per cent carbon dioxide reductions (including those relying on nuclear power or the other currently contentious technology: carbon capture and storage, CCS) will inevitably have societal consequences. For instance, in order to achieve such drastic CO_2 reductions the changes to transport and mobility must be substantial. How will society constrain the behaviours of both motorists and the transport industry in order to deliver the changes required? Without the deployment of the contentious technologies of nuclear power and CCS the required reductions in carbon dioxide emissions would appear to be more expensive (Marsh et al., 2003; DTI, 2003). It is not the purpose of this chapter to tackle the tricky economics of nuclear power or of carbon capture and storage.[2] Rather this chapter seeks to assess whether achieving a 60 per cent CO_2 reduction without CCS and nuclear power would necessitate uncomfortable lifestyle changes affecting many of the more enjoyable experiences of modern life. If the measures to achieve climate stability are draconian, then the kick in the small of your back when you hit the accelerator in your car could in future become a distant memory as vehicle design alters to improve efficiency and eliminate wasteful excess torque. Also air conditioning could return to being a rare luxury in the United Kingdom. Many people could object to being forced to pay for mitigation services, such as CCS which represent a new cost in the system, which they do not desire and for which they cannot see a direct need. It is issues of this type that have the potential to arouse public anger and to alter public attitudes to nuclear power. The growth of fly-tipping in the United Kingdom in recent years (following moves to extract fees for waste disposal) could be an example of the kinds of societal tensions that can result from forceful moves in environmental policy. In this case it is arguable that problems arose despite the fact that the majority can be expected to support the policy. It is precisely when the will of the majority is perceived to be attacking the rights and privileges of a minority that the strongest political tensions can occur. While there is clearly no perceived right to fly-tip, and there is little or no majority sympathy with such illegal minority behaviour, there is clearly much frustration around the issue, both with the fly-tippers and for those saddened by the damage to the countryside. Another example of potential relevance is that in the United Kingdom there is a minority opinion that individuals have the right to hunt foxes with hounds. The recent anger of this minority at the perceived loss of a key part of their way of life (as a result of the Hunting Act 2004) is both powerful and visible. If measures to achieve 60 per cent carbon dioxide reductions are advanced without a return to nuclear power and without the development of carbon capture and storage then there would appear to be an enhanced risk that draconian and politically unpleasant policies might be required to stabilise the climate. It is perhaps not unimaginable that in the future lovers of classic twentieth-century sports cars might unite with those with an affection for a traditional coal fire, or for air conditioning, and find common cause to oppose the green authoritarians.[3] It is not impossible to imagine an energy policy backlash not unlike the emergence of the pro-fox hunting group the Countryside Alliance. In fact one might argue that a related backlash has already occurred in continental Europe and the United Kingdom with the fuel price protests of late 2000.[4] Earlier this chapter posited the idea that public nervousness with nuclear power might be related to a perceived fear that nuclear power represents a threat to liberal society. As the threat of climate change looms ever larger there is perhaps the possibility that public attitudes might swing in favour of nuclear power in an attempt to avoid the prospect of even more authoritarian policies. If the future of nuclear power does rest upon a balance of such fears it is clearly in the interests of the nuclear industry to move away from traditional technocratic approaches. It would appear possible to develop scenarios for nuclear power that allow it to help reconcile energy policy with continued liberal democracy while simultaneously assisting the world to reduce drastically its carbon emissions.

A new paradigm for nuclear power?

This chapter considers the possibility that the nuclear power industry might move towards democratic multi-stakeholder processes and decision-making. In such a future the details of the industry itself must adjust substantially from those developed over the last 60 years under a technocratic paradigm. In order to appreciate the issues underpinning such shifts it is necessary to consider in some detail issues of risk and the public perception of risk.

Michael Mehta argues that in order to make progress on technology and risk one must first consider risk to be a socially constructed concept (Mehta, 2005). This author would not go so far, but rather would argue that there are two distinct concepts to be considered. First there is true 'risk' – ideally an objective quantitative reality and often interpreted via mathematical models and constructs. Various definitions of 'risk' are used in the literature, but each relies on probability and quantitative assessment. Our intention here is to consider a separate concern – the human response to risk. This response or attitude is indeed a social construct. Of risk and risk perception, it is the former that has thus far dominated technocratic decision-making in nuclear power, but it is the latter that will, and should, more strongly determine the shape of any nuclear renaissance.

Nuclear energy is not the only technology and policy issue that is likely to be shaped more by public perceptions of risks than by considerations of risk itself. One clear example is the case of genetically modified (GM) crops in Europe. Those deploying GM technologies, or for that matter nano-technology, may have much to learn from the nuclear energy experience.

This author has argued previously that for 50 years the nuclear industry has heard that the public is scared of the dangers of nuclear power and in response the nuclear industry has worked to minimise the dangers (Nuttall, 2005a, p. 113). A radical shift from technocratic leadership to more democratic processes would not now be so pressing an issue if the industry had worked from the start to minimise fear as hard as it has worked to minimise danger. If the nuclear industry is to find a future associated with lower levels of public fear then it must first better appreciate the sources of such anxiety. Such thinking takes the industry firmly into the domain of socially constructed public perceptions and away from the world of quantitative or 'true' risk.

Peter M. Sandman has provided numerous provocative insights into these matters through his suggestion that for practical purposes *risk equals hazard plus outrage*. Hazard corresponds to 'true risk' as described above, while 'outrage' refers to the social response (fear, anger, etc.) (Sandman, 1993). In Sandman's terms therefore this chapter argues that, in the case of nuclear power, the industry should have done more to recognise, understand and address the outrage rather than simply focussing upon minimising the hazard.

In a paper examining issues facing those planning to engage in public communication about risk Jill Meara reports on a British Department of Health study on the fright factors for risk (Meara, 2002).

Presenting a list similar to one used by Sandman, Meara notes that risks are less acceptable and more feared if they are perceived to be

- Involuntary
- Inequitably distributed in society
- Inescapable
- Coming from an unfamiliar or novel source
- Causing hidden or irreversible damage particularly dangerous to children or future generations
- Causing dreaded illnesses (e.g. cancer)
- Poorly understood by science
- The subject of contradictory statements from scientists in authority.

Nuclear power is remarkable in that it exhibits, or is perceived to exhibit, all of the fear factors listed above. However, it is possible to conceive of a nuclear power system designed to reduce the impact of some of the fear factors listed. In the United Kingdom these fear factors have traditionally had little or no influence on policy for nuclear energy.

National differences in the politics of nuclear energy

The country that has experienced the most incendiary nuclear energy politics is Germany. It is interesting to speculate that this tension is a direct consequence of Germany's totalitarian past and its front line role during the Cold War. Werner von Lensa has characterised the German nuclear energy policy experience as a 'quasi-religious war' suffering unduly from a polarised and dualistic approach to the issues (von Lensa, 1998).

With these considerations in mind it is helpful to consider Scandinavian developments in the nuclear fuel cycle. Among the older professionals in the European nuclear industry Sweden is still thought of as a country where policy for nuclear power became derailed by misplaced environmentalism in the 1970s. In fact the nuclear power sector in Sweden functions well to this day. While Sweden has just shut down its oldest nuclear power plant (Barseback-2) a programme of modernisation and capacity improvement at its other nuclear power plants will ensure that, in the short term at least, nuclear electricity generation in Sweden will increase.[5] The Swedish progressive thinking of the 1970s has however led to a remarkably positive current position for nuclear power in that country. First Sweden took a clear decision against reprocessing on the grounds that it did not want an inventory of separated civil plutonium. Given the relatively low price of uranium and the growing concerns for nuclear safeguards and security measures, Sweden's plutonium decision appears to have been the right one. As such the nuclear waste inventory in Sweden consists of spent fuel. Another remarkably prescient decision was that the spent fuel should be stored in a specially designed facility known as the CLAB built many metres underground in excavated granite caverns (Wikstrom, 1998). This approach differs from practice in several other European countries where similar materials are stored in surface facilities. Following the terrorist attacks on the United States on 11 September 2001 the Swedish decision to store spent nuclear fuel underground seems to have been wise. Lastly the Swedes and the Finns have been making good progress towards the very long-term management of waste spent fuel. Sweden has constructed an underground rock laboratory at Åspo near Oskarshamn. The successful completion of this facility contrasts remarkably with the 1997 failure of Nirex in the United Kingdom to receive planning permission for a similar facility known as the Rock Characterisation Facility. In the context of this chapter, however, perhaps the most important aspect of the Åspo facility is its surface architecture. In marked contrast to nuclear facilities, such as Areva's La Hague reprocessing facility near Cherbourg, France with its brutal box-like buildings and its spiky antennas and towers, the Swedish Åspo facility is reminiscent of a quaint Scandinavian building in a nautical tradition (note the widow's walk) and also with a slightly agricultural impression (see Figure 16.1).

Figure 16.1   Surface buildings of the SKB Åspo Underground Rock Laboratory for radioactive waste management research near Oskarshamn, Sweden (*Source*: SKB)

The architecture appears to have been determined by a conscious attempt to minimise fear through familiarity and positive association in an area with proud heritage in both fishing and farming. Some people might regard this approach as including an unethical attempt to deceive. To this author's impression, however, such arguments merely reveal a lack of understanding of the history of architecture. Over the centuries each new structural function has looked to antecedents for architectural inspiration. Many of the first mills and factories of the British Industrial Revolution of the late eighteenth century were constructed with forms reminiscent of Palladian classical architecture. In such a spirit there would appear to be nothing deceptive or dishonest in the surface structures of the Åspo facility being constructed to look like other buildings characteristic of the local landscape.

Towards transparency and inclusion

Across the Baltic Sea other moves towards the democratisation of nuclear power have been occurring. For instance, Finland was the first country in Europe to announce new nuclear power plant construction and in so doing forms the vanguard of the nuclear renaissance. Finland also finds itself in a leading position in respect of policy for radioactive waste management. From 1983 to the present Finland has made steady progress towards the construction of a repository at Olkiluoto (Nuttall, 2005a). Finnish progress has been made on the basis of community volunteerism, transparency and mutual engagement between the local community and policy makers. Trust is key to the Finnish model with the nuclear waste policy makers trusting the local community by providing them with a community veto throughout the lengthy process and a reciprocal trust by the community of the policy makers that the facility is indeed as safe as it has been described. It is arguable that such processes of joint community and expert decision-making works best in a Scandinavian cultural and societal setting. Given the 1997 collapse of the plans by Nirex for the Rock Characterisation Facility (RCF) at Sellafield in Cumbria, it would appear that the Scandinavian approach is now worth trying in the United Kingdom. Malcolm Grimston and Peter Beck have described the original Nirex strategy with its stakeholder communication placed towards the end of the process as having been one of 'Decide, Announce, Defend and Abandon' (Grimston and Beck, 2002). In the United Kingdom and since the election of the Labour government in 1997 there have been significant moves in the United Kingdom towards more democratic processes for policy developments in radioactive waste management. For instance a new semi-expert policy development body was constituted in 2003 and tasked with reporting in 2006: the Committee on Radioactive Waste Management (CoRWM). CoRWM has attempted to operate in a transparent way and to be receptive to novel thinking. Nevertheless, during its deliberations CoRWM suffered from tensions arising from its requirement to balance sociological concerns with more traditional technical matters.[6] Transparency is a concept that underpinned CoRWM's work and it is also a lesson learned by Nirex following the failure of the RCF.

In August 2002 Nirex published a transparency policy learning lessons from the RCF experience (Nirex, 2002).[7] Some items of confidential information from the past, however, remained secret after the launch of the transparency policy. In particular the matter of greatest concern was a secret list of ten sites considered by Nirex for intermediate level waste disposal in its (now completely ended) original research programme. The reason given previously for retaining secrecy of this information has been that it would cause blight on properties known to be near these sites. The process leading to that site list is, however, now completely ended and it seems likely that the old site list is of no future relevance for radioactive waste policy which is starting from scratch in the United Kingdom. For that reason Nirex agreed in 2005 to release the information under the terms of the UK Freedom of Information Act.[8]

Originally constituted as a creature of the nuclear industry, Nirex reported in 2005[9]:

Nirex has this year (1 April 2005) been made independent of the nuclear industry, in a move that will boost transparency and accountability in the long-term management of radioactive waste. Independence for Nirex means that the company, set up in 1982 to implement a strategy for the safe disposal of wastes of low and intermediate-level radioactivity, can take the first step towards making a real and legitimate contribution to the Government's objective of implementing a long-term strategy for managing radioactive waste.

There would appear to be the possibility for progress towards a more democratic nuclear energy system by means of a greater concern for local community support. Recent United States experience in radioactive waste policy reminds us of another model for 'democratic' decision-making – publicly endorsed strong central leadership (Grimston, 2005). Since the events of 11 September 2001 the United States Federal Government has pushed forward policy for a national permanent waste repository at Yucca Mountain in Nevada. These measures, however, are being hotly contested by the state of Nevada through the courts and it is not yet certain whether the United States Government's use of strong Federal authority, backed by national democratic mandate will prevail. The Finnish experience of the politics of consensus would appear to be a more successful model for policy progress than the US model of national democratic structures over-riding the will of local people. There are numerous differences between the United States, Finland and the United Kingdom and any, or all, of them might limit the transferability of approaches between countries. For instance the countries differ in their constitutions with differing levels of central authority, they differ in geographical size and population, they differ in the level of social homogeneity and cohesion and, of course, they have different historical legacies. Nevertheless, given the failure of more domestic approaches it would appear timely for the United Kingdom, in particular, to seek to learn from international experience.

Publicly accepted and safe enough

So far in this chapter it has been argued that nuclear power can and must become ordinary and that the decisions driving the future of the industry should be shaped by the opinions of the widest possible community of local stakeholders. It is worth policy makers examining the possibility that such democratic processes would indeed yield a more sustainable com-mercial nuclear power industry. Possible measures consistent with lower public anxiety and greater public consensus include the monitored retriev-ability of nuclear wastes in deep underground repositories rather than the originally more orthodox, and marginally safer, approach of deep under-ground disposal with the facility closed with a backfill of bentonite[10] clay or concrete. Such an approach would increase the chance of public accept-ance at the price of a small, but acceptable, erosion of safety. Here it is argued that if the preferred approach of the public is safe enough, then it should be adopted. Not all technologies are safe enough however. Some technologies, such as the disposal of radioactive wastes in outer space, while receiving relatively high levels of public interest, are regarded by most experts as being unacceptably dangerous.[11] The option of firing radioactive wastes into space must be rejected as it is simply not 'safe enough'. Together with others this author has written previously in sup-port of greater levels of investigation into the partitioning and transmu-tation of radioactive wastes (Nuttall et al., 2005b).[12] It would appear that such approaches are particularly interesting as they exhibit relatively high levels of public support combined with relatively high levels of expert con-cern as to safety. It is important to note that in respect to more democratic nuclear fuel cycles, concern for the environment and for safety may actually reside more strongly with the experts than with the public. It would be an unusual situation for nuclear power if its safety became one of those areas of technology policy where the more you know the more you worry. In a move to a more democratic nuclear fuel cycles there are risks of such a situation developing and therefore experts must always be vigilant that their industry is indeed safe enough.

This author has argued previously that the nuclear industry's extreme safety culture, in which the lives of nuclear workers are to be protected as a first priority, can actually erode public sympathy (Nuttall, 2005a). For reasons discussed earlier the public are actually quite accepting of informed and appropriately remunerated nuclear workers risking their lives in an industrial setting. Similar social contracts exist in many industries such as fossil fuel extraction and civil engineering. What the public particularly resents is an imposed risk falling on relatively ignorant members of the public. Clearly, when it comes to the politics of deploying hazardous technologies, not all deaths are equal. The rational nuclear industry view that the deaths of 'real people' are more important than an equal number of deaths of unknown and unknowable people in the distant future, runs somewhat counter to public perceptions of these issues. The technocratic view is that the known deaths of identifiable workers are clearly preventable and as much as possible must be done to minimise such events. The vanishingly remote risks to large numbers of current and future members of the public simply cannot be handled in the same way. All must be done to reduce those risks, but it is not done via the same procedures as worker safety. Such disconnects between the treatment of worker safety and public risks can be a source of public concern. Policy progress can be made, but the nuclear industry must be careful to avoid the perception that it protects its own above all else. A move from technocracy to democracy can only help in this regard.

When the technocrats of the nuclear fuel cycle turn their attentions to other stakeholders they still often take the view that education is the key to greater public acceptance. Their reasoning is such that they believe if only the public could come to know what they know, then the public too would share the expert perspective and agree with the expert conclusions. This view is known as the 'deficit model' and it is widely acknowledged to be flawed. Sandman critiques it well in Chapter 3 of *Responding to Community Outrage*. He argues that while it is necessary to minimise the hazard and importantly to explain the hazard to concerned public, such measures are usually insufficient in the absence of separate efforts to minimise the outrage. Both effective communication and real risk minimisation must go hand in hand. He states it even more straightforwardly when he says: 'Risk communication that is deployed as a substitute for risk reduction is doomed to fail and rightly so.'

One's attitude to power and control is a fundamental emotional and political thought and, as such, it would be foolish to assume that such social attributes of the individual are easily altered by education. Neither the public nor nuclear industry professionals are exempt from these realities. It would appear, therefore, that the best strategy for the nuclear industry is not to educate the public into membership of the technocracy, but rather for the technocrats to listen to and to be more led by the public. In so doing they might seek to become truly ordinary members of the polity. Such thinking leads us to the domain of Brian Wynne and other proponents of the *contextualist* perspective on public attitudes to science and technology. Wynne stresses that science itself is socially negotiated (Irwin and Wynne, 2004). This chapter has argued that nuclear energy has had low levels of such social contextualisation. A more contextualist approach would allow any nuclear renaissance to be built upon more democratic foundations. The chapter has noted that true probabilistic risk is not a social construct. Furthermore, it is important to note that the use of nuclear fission to generate electricity is clearly not simply a social construct. This author is reminded of the late Keith Pavitt's resonant aphorism that *no-one ever flew the Atlantic on a social construct* (Pavitt, 1998). However, this chapter accepts that public attitudes to nuclear power are socially constructed and that these attitudes, provided that the resulting policy implementation is safe enough, should properly have a role in shaping policy for nuclear power.

In calling for greater levels of democratic leadership in nuclear power decision-making it is important to stress that public opinion is not confused with the opinion of pressure groups and non-governmental organisations. Such bodies are important stakeholders to decision-making, but this chapter draws a firm distinction between such attitudes and those of the general public. It is the public voice that this chapter seeks to amplify, not the lobbying of single-issue pressure groups.

This chapter concentrates on the premise that public acceptance will be key to the future of nuclear power. Polling by MORI (Figure 16.2) illustrates that recently the proportion of the British public with a positive opinion about nuclear power has started to exceed those with a negative opinion. Even more importantly, however, roughly half the British public have no real opinion (Knight, 2005). This chapter does not argue that if the economic and environmental benefits of nuclear power are real then policy makers should seek to persuade the public to accept the nuclear option.

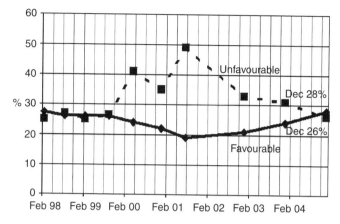

Figure 16.2   MORI: All Great Britain general public polling of public attitudes to the nuclear industry (Base MORI Omnibus polling. Approximately_2000 face-to-face interviews of the general public aged 15 and over at 210 sampling points) (Knight, 2005)

Rather it is suggested here that public attitudes must be a component of the policy process from the start. An open and transparent approach is to be preferred as a bulwark against authoritarianism. The MORI data tell us that as we enter a period of potential nuclear renaissance we must not just accommodate the views of those with strong opinions, but also recognise that many in the British population do not, at present, care very much.

Conclusions

Nuclear power has many beneficial attributes that motivate us to consider it as an important contribution to future global energy supply. In order to play such a role this chapter suggests that it is important that policy and decision-making for nuclear power is carried out in new and more inclusive ways. Nuclear power must move fully to a paradigm characterised by democracy and consensus. In this author's opinion a nuclear renaissance in Western Europe is only possible if founded upon principles of informed consent and stakeholder-based decision-making. The nuclear industry that results from more socially constructed processes may not be quite as safe and may be somewhat more expensive than that suggested by the technocratic experts, but within reason such concessions are both appropriate and proper. If such a democratic future for nuclear power will be safe enough, economically affordable and environmentally benign then this author recommends that policy makers support its development. Indeed, if nuclear power is to endure, the coming nuclear renaissance must be accompanied by a *nuclear enlightenment.*

Acknowledgements

The author is most grateful to Robin Grimes, Malcolm Grimston, Samantha King, Werner von Lensa, David Hamilton and Waclaw Gudowski and to an anonymous referee from the University of Cambridge, Electricity Policy Research Group for insightful and helpful comments. I am also most grateful to Simon Smith for suggestion that I follow my work on *nuclear renaissance* with consideration of a *nuclear enlightenment.* I am grateful to SKB, Sweden for making available the image used for Figure 16.1. All errors and unattributed opinions expressed in this chapter are those of the author alone and, as such, no responsibility for such matters rests with those that have kindly provided assistance.

Notes

1. In 1984 linguist Thomas A. Sebeok was tasked by the US office of Nuclear Waste Isolation to find a way in which to convey a warning message about the dangers of a nuclear waste repository in a way that would be resilient for 10,000 years or 300 generations. Sebeok concluded that over such long periods both languages and the contexts of languages vanish. His controversial suggestion was the construction of an 'Atomic Priesthood' capable of sustaining the truth from generation to generation and positioned to warn intruders of the dangers of any curiosity.
2. Readers with an interest in the economics of nuclear power are recommended to consult the 2003 MIT report *Future of Nuclear Power* or University of Chicago report of August 2004, *The Economic Future of Nuclear Power.*
3. There are parallels with the debate over 'eco-imperialism' concerning the relationship between first world environmental non-governmental organisations and developing countries. See, for instance, Paul K. Driessen's controversial book *Eco-Imperialism, Green Power Black Death* (Driessen, 2003).
4. See http://news.bbc.co.uk/1/hi/uk/924574.stm – accessed June 2005.
5. See http://msnbc.msn.com/id/8058171/ – accessed June 2005.
6. See for instance: http://www.timesonline.co.uk/article/0,,2-1638937,00.html – accessed June 2005.
7. Issues concerning nuclear waste policy and management are discussed elsewhere in this volume.
8. See: http://www.nirex.co.uk/index/inews.htm – accessed June 2005.
9. Source: Nirex website: http://www.nirex.co.uk/index/iabout.htm – accessed June 2005.
10. Some backfill strategies involving bentonite clay are in principle retrievable, but would require significant effort.

11. The use of plutonium-fuelled radioisotope thermoelectric generators on spacecraft such as the Cassini probe notwithstanding (see: http://www.seds.org/spaceviews/cassini/rtgpages.html – accessed March 2006).
12. Partitioning is the separation of radioactive waste into chemically more homogeneous streams. Transmutation is the use of nuclear physics techniques to convert harmful radioactive isotopes into shorter lived or more benign material.

References

Department of Trade and Industry (DTI), UK (2003) *Energy White Paper – Our Energy Future – creating a low carbon economy*, London.

Driessen, P.K. (2003) *Eco-Imperialism Green Power Black Death*. Merril Press, Bellevue, WA, USA.

Grimston, M.C. (2005), private communication.

Grimston, M.C. and P. Beck, (2002) *Double or Quits – the global future of civil nuclear energy*. Royal Institute of International Affairs, Earthscan, London.

Hore-Lacey, I. (2003) *Nuclear Electricity*, 7th Edition. Uranium Information Centre Ltd and World Nuclear Association, www.uic.com.au/ne.htm, accessed 03 October 2006.

Irwin, A. and B. Wynne (2004) *Introduction* in *Misunderstanding Science? The Public Reconstruction of Science and Technology* (A. Irwin and B. Wynne, eds). Cambridge University Press, Cambridge, UK, June 2004.

Jungk, R. (1979) *The Nuclear State*, Translated by Eric Mosbacher from *Der Atomstaat*, John Calder (publishers) Ltd, London.

Knight, R. (2005) 'What do the polls tell us?' *Nuclear Engineering International*, April, pp. 24–25.

MacKerron, G. (2004) Nuclear power and the characteristics of ordinariness – the case of UK energy policy. *Energy Policy* **32**, 1957–1965.

Mander, J. (1978) *Four Arguments for the Elimination of Television*. William Morrow, New York, p. 44.

Marsh, G., P. Taylor, D. Anderson, M. Leach and R. Gross (2003) *Options for a Low Carbon Future phase 2*, Future Energy Solutions, AEA Technology, http://www.dti.gov.uk/energy/whitepaper/phase2.pdf as of June 2005, February 2003.

May, M. and T. Isaacs (2004) Stronger measures needed to prevent proliferation. *Issues in Science and Technology*, **20**(3), 61–69, Spring 2004.

Meara, J. (2002) Getting the message across: is communicating risk to the public worth it? *Journal of Radiological Protection* **22**, 79–85.

Mehta, M.D. (2005) *Risky Business*. Lexington Books, Oxford, p. 14.

Nirex (2002) *Transparency Policy*. UK Nirex Ltd, Harwell, Oxfordshire, England.

Nuttall, W.J. (2005a) *Nuclear Renaissance – technologies and policies for the future of nuclear power*, IOP Publishing, Bristol.

Nuttall, W.J. (2005b) Potential for British research into the transmutation of radioactive wastes and problematic nuclear materials, D.G. Ireland, J.S. Al-Khalili, W. Gelletly, *Int. J. of Critical Infrastructures* **1**(4), 380–393.

Pavitt, K. (1998), private communication.

Rothwell, G. and B. van der Zwaan (2003) Are light water reactor energy systems sustainable? *J. Energy and Development* **29** (1), 65–79.

Royal Commission on Environmental Pollution (RCEP) (2000) 'Energy The Changing Climate', 22nd Report, London.

Sandman, P.M. (1993) *Strategies for Effective Risk Communication*. American Industrial Hygiene Association, Fairfax, VA, USA.

von Lensa, W. (1998) *Sustainability and Acceptence – New Challenges for Nuclear Energy*, In: Proceedings of a Technical Committee Meeting Held in Beijing, People's Republic of China, 2–4 November 1998; International Atomic Energy Agency, International Working Group on Gas-Cooled Reactors, Vienna, Austria. IAEA-TECDOC–1210, pp. 237–246.

Wikstrom, M. (1998) *Radioactive Waste Management in Sweden: experience and plans*. Presented at Int. Symp. Storage Spent Fuel Power Reactors, Vienna, Austria (Available at: http://www.skb.se/upload/publications/pdf/wikstrom-cambridge-98.pdf January 2006).

Winner, L. (1986) *The Whale and the Reactor*. University of Chicago Press, Chicago. pp.19–39.

17

UK Energy Choices:
An Enlightened Future?

David Elliott

The previous chapters have ranged widely over various aspects of the debate over the future roles of nuclear and renewables and associated specific problems and issues. There is no way that the various conflicting views can be resolved in this book, but I have tried to summarise some of the basic issues in the Table 17.1. These assessments are inevitably couched in general terms, but should provide a reasonably acceptable framework for comparison of merits and demerits.

Obviously there will be disagreements about specifics, and, perhaps more importantly, about which of the categories is most important in terms of making choices. In addition, the choice we face is clearly not simply that between nuclear and renewables. Perhaps more important than the debate around the specific issues summarised in Table 17.1 is the structure and dynamics of the wider debate around policy responses to climate change. In particular – what is the best mix of technologies for a sustainable future and how should they be integrated together?

For example, there is an ongoing debate about how the overall energy system should be developed. In recent years, the emphasis has increasingly been on smaller-scaled plants – combined cycle gas turbine plants and wind farms of the order of 20–100 megawatts instead of giant gigawatt coal and nuclear plants. We are moving from a system in which giant plants send power to users down long grid lines, to one in which smaller plants are embedded in more localised grid networks. There are economies of scale with big plants, but also losses due to transmitting power over long distances. The trend is towards decentralisation – including generation by consumers themselves, using domestic-scale micro-power systems. In that context, big nuclear plants may be out of place.

In addition there is the key issue of the role of energy efficiency – if energy wastage can be avoided then it becomes easier to meet the reduced

Table 17.1   Comparison of nuclear and renewables

	Nuclear	Renewables
Resource lifetime	Uranium reserves ~100 years *at current use rates* ~1000 years with FBR?	Effectively infinite resource lifetime
Resource scale	Currently ~6% of world energy, ~17% of electricity Could perhaps be doubled? Or trebled? i.e. to 50% of world *electricity*. But lifetime of the resource would then be limited	Currently ~6% (with hydro), ~17% of electricity Projection: 50% of world *energy* by 2050 (RE2004/Bonn Conference)
Eco impacts	Infrastructure impacts, cooling water impacts, risks from very long-term wastes ~10,000 years	Local visual intrusion and land use conflicts, some local eco-impacts (especially with biomass and large hydro)
Safety	Major accidents ~10,000 deaths, occasional/routine emissions ~100's of deaths	Generally low risk, except large hydro ~10,000 deaths
Costs	High and could rise as uranium resource dwindles, but new technology could emerge	Some high, but most are moderate, and all are falling as technology develops
Output	Electricity only, but could be used for direct heat or hydrogen production	Diverse sources: electricity, heat, fuels
Reliability	Occasional shut downs	Some rely on variable sources so need grid integration to balance outputs
Supply security	Uranium deposits limited to a few locations	Widely diffused energy sources
Security risks	Significant terrorist targets, plutonium proliferation threat	No significant problems except with large hydro

energy demand through whatever means is chosen. Most people accept that energy efficiency and demand side strategies are crucial, although the level of commitment to this varies. In the past, the nuclear lobby has sometimes tended to be dismissive of what could be achieved by energy conservation – especially when this was promoted as a solution by environmental pressure groups. The renewables' lobby, which has emerged in part from the 'green' end of the spectrum, has usually been more supportive of energy conservation, not least since if demand could be constrained, then it would be more credible to meet it with renewables. At the same time, there is still a strong conviction that, although conservation and efficiency can and should achieve a lot, the main issue is on the supply side. However much we avoid energy waste, we will still need new energy supplies, and if these are to be carbon free, then the only options at present, apart from sequestration/carbon capture and storage (CCS), are nuclear or renewables.

While nuclear proponents are clearly convinced that their case for playing a major role on the supply side is robust, it also seems clear that there is still a long way to go, even for the most enthusiastic supporters of nuclear. Not least since renewables are coming on rapidly as arguably a much more attractive set of options – along possibly with carbon sequestration and of course energy efficiency. So we are left, on one hand, with the hope by the nuclear proponents that better times will come based on new technology, coupled with a desire to be seen to be supportive of renewables, as allies rather than rivals; and on the other, with the belief by many renewables enthusiasts that their favoured technologies could and should provide the main way forward – with nuclear being seen as a threat to their development.

There may be some potential common ground in the idea of having both nuclear and renewables, but for the moment most of the enthusiasm for this compromise seems to be from the nuclear side. Few enthusiasts for renewables believe that major nuclear and renewables programmes could be expected to co-exist peacefully. As I argued in my chapter, and as Mitchell and Woodman also argued in Chapter 11, there are certainly risks that revived support for nuclear could undermine continued support for renewables. Some of these rivalries may be petty and not based on substantial conflicts, although there are also potential strategic disagreements. These are not limited just to the renewables community. It is interesting, for example, that some parts of the nuclear fission lobby are sometimes surprisingly dismissive of nuclear fusion, which it evidently sees as at best a long-term option, but one which might deflect funding from the fission programme in the short term. The renewables lobby has similar views – hard pressed developers resent the large-scale funding of fusion. Both the nuclear and renewables lobbies might also be expected to be a little concerned about the potential for deflection of resources into clean coal/carbon capture and storage.

These are conflicts essentially over limited resources, although they are also conflicts over which mix of technologies is seen as the most viable for dealing with climate change. If fears about climate change grow, then the pressure to resolve these disputes will also grow. However, a simple policy of 'lets have more of everything', in effect pushing every panic button in the control room, may not be the best way forward.

While this strategic debate will no doubt continue among experts, policy makers and planners, there is also the wider public debate. The renewables lobby is relatively weak in industrial and even political terms, but it does have wide support from environmental pressure groups and the public. By contrast, the nuclear lobby is relatively well resourced and influential in political circles, with considerable backing from the establishment organisations – from the CBI to the various professional institutions. Its weak point, at least until recently, has been the lack of support from the public and the hostility of most 'green' groups. It could of course be that fears about the scale of climate change will lead more 'green' minded people to embrace nuclear, as well as renewables. Fears about energy security might have a similar effect. However, any such shift will probably depend on the scale of the nuclear programme. A simple replacement programme may be easier to sell than a major expansion, not least since the former could make use of existing sites where local opposition may be less. Given this situation, it is probably unwise for nuclear proponents to talk, as some have, of a major programme of 20 new reactors supplying 30–40% of UK electricity.

On the other side of the debate, the renewables lobby has sometimes overstated its case by alluding to large potential contributions that it claims could be made. The reality, at least so far in the United Kingdom, is not too impressive. To be fair however, the situation elsewhere, notably in Germany and Denmark, is very different, illustrating what could be done in the United Kingdom, which after all has a much larger and better renewable resource base. It is perhaps then not surprising that some UK based renewable energy enthusiasts feel thwarted by what they see as a lack of serious commitment by government to renewables. On this view, it is not that renewables have been 'tried and failed' – they have not really been given chance to show what they can do. And so we have gone back to nuclear by default.

Perhaps the key issue is the respective public images of the two technologies. Both strive to project themselves as modern and future orientated, and yet both are relatively old – nuclear has been around 40 years or more and some renewables are based on pre-industrial practices. Nuclear does have the edge, in that it is the result of scientific breakthroughs before and during the Second World War, but in engineering terms it might be argued that using heat from uranium fission to raise steam for turbines which can convert at best 35% of the input energy into electricity is a pretty inelegant approach for the future. Equally it could be argued that wind and water mills were abandoned in the past since they were ineffi- cient and that modern variants are not that much better. However, mod- ern wind turbines, wave energy devices and tidal current turbines have a different kind of appeal – what they do and how they work is clear and transparent. There are no hidden downsides – what you see (visual intru- sion in the case of wind turbines) is what you get. By comparison, it is hard for most people not to see nuclear plants as to some degree sinister and menacing.

In Chapter 16, Nuttall calls for a new enlightenment to help sustain a nuclear renaissance. However it seems unlikely that, even given wide- spread public consultation and stakeholder dialogues, nuclear will ever be anything less than grudgingly accepted in the United Kingdom. In part this may be because of the past history of secrecy and overly posi- tive and even arrogant assertions about safety and cost. But it is also per- haps due to the very nature of the technology – it is complex and mysterious. Wonderful perhaps to some technophiles, but also perturb- ing to those who look for simplicity and elegance and find it in, for example, modern photovoltaic solar devices. Somehow, in the final analysis, relying on solar photons seems wiser than relying on nuclear neutrons.

Abstract ideas like this will of course not be uppermost in many peoples minds when they come to respond to specific projects, whether nuclear or otherwise. As wind energy developers have found, straightforward 'NIMBY' issues relating to visual intrusion, and alleged impacts on house prices, have more influence. The renewables lobby would no doubt share the nuclear lobby's conviction that what is needed is a new enlighten- ment, so that NIMBY opposition could be replaced by an awareness of the need to balance local impacts against global gains. However, it remains an open question as to whether the ecological enlightenment that the renewables community looks for is the same as that sought by the nuclear lobby. After all, quite apart from technical disagreements, there are differ- ent views as to what makes for a desirable future, and on how we should seek to attain it.

The dominant technocratic view has been challenged by radicals of various kinds over the years, and to some extent nuclear power is seen by its critics as inseparable from the future based on centralised control and continued economic growth of the current type. Wide acceptance and effective practice of the newly emerging 'green' world view will probably require a change in social and political perspectives in which centralised technologies and economies may have only a limited relevance.

The debate over nuclear is thus wider than just the debate over specific impacts and costs, which is perhaps why the nuclear lobby sometimes characterises its critics as 'ideological'. For the nuclear lobby, technology is neutral, while, as is noted in this book, some 'greens' believe that nuclear technology is inherently 'ideologically' flawed, although others are beginning to question this belief. There are probably no absolute answers to such questions – other than saying it depends on the social and political context. That sort of relativism worries some people. It smacks of the belief at one time heard on the Left that nuclear power would be acceptable 'under socialism', but not under capitalism. We live in ostensibly less ideological times, but as this book attempts to show, that does not mean that there are not contested views about which technologies are best suited to ensuring a sustainable future and as to what form that society should take. To the extent that society and technology interact to shape each other, it is vital that this debate is open to a wide range of influences, as part of a wider process of social engagement. However, what the eventual outcome of that will be is unclear – whether it would result in nuclear power becoming accepted as green technology remains to be seen.

How quickly do we need to resolve this issue? We are sometimes told that there is a looming energy gap and that, to deal with climate change we must find new carbon-free sources rapidly. It is true that we need carbon-free sources, but we are fortunate in that there are plenty of them at various stages of development. There is certainly a need for urgent action, but we should not be panicked into making hasty and potentially irreversible decisions. It could be argued that it was unfortunate that the 2006 UK Energy Review was to some extent side-tracked into the nuclear debate, with the implication being that we could retreat back to the traditional 'big centralised supply' approach. Instead we need to think carefully about what sort of energy system we want in the future – and which technologies are best for it, adopting a wider more comprehensive approach. This book has attempted to set the nuclear debate in a wider context, but inevitably the main focus has been on nuclear issues. Other books in this series, and in particular *Sustainable Energy*, attempt to extend the wider approach.

What next in the UK? A 2009 update

Although alternative views are presented, much of the analysis in this book has suggested that new nuclear is not the best option for the UK. However, the UK government clearly feels that nuclear power could and should play a key role, as witness the conclusions of the Energy Review published as the first edition of this book went to press in July 2006. It claimed that 'Based on a range of plausible scenarios, the economics of nuclear now look more positive than at the time of the 2003 Energy White paper' and concluded that 'nuclear should have a role to play in the future of the UK generating mix, alongside other low carbon-generating options'.

Subsequently, as noted in the (updated) Introduction to this edition, a private sector led programme has begun to unfold, with background support, but no direct financing, from the government. However, in parallel, a new expanded renewables programme is also emerging. There have already been some collisions between these two energy options. For example, one of the 11 sites selected as possible locations for nuclear plants has a working wind farm established on it, which it was suggested would have to be removed if the nuclear plan went ahead. A perhaps more strategic sign of the way industrial priorities might change was the decision by the major engineering company AMEC to withdraw from its wind farm work and focus on nuclear power, as part of a new consortium taking over Sellafield's waste management activities.

Operational conflicts also seem possible. EDF, the French company that now owns British Energy, has proposed building several European Pressurised-water Reactors (EPRs) in the UK, but in its submission to the UK government's 2008 consultation on the 'UK Renewable Energy Strategy' it warned that 'As the intermittent renewable capacity approaches the Government's 32% proposed target, if wind is not to be constrained (in order to meet the renewable target), it would be necessary to attempt to constrain nuclear more than is practicable'.

It explained that 'EPR nuclear plant design can provide levels of flexibility that are comparable to other large thermal plant. However, there are constraints on this flexibility (as there are for other thermal plant). For example, the EPR can ramp up at 5% of its maximum output per minute, but this is from 25% to 100% capacity and is limited to a maximum of 2 cycles per day and 100 cycles a year. Higher levels of cycling are possible but this is limited to 60% to 100% of capacity.'

The point is that it will be hard to have large numbers of relatively inflexible nuclear plants and also large contributions from variable renewables like wind on the grid. In the absence of significant energy storage or power export facilities, at times, for example at night in summer, when electricity demand is low, some of the power from one or the other, or both, will have to be wasted.

EDF's choice as to which should take priority, at least in the electricity sector, is clear: 'A lower volume of intermittent renewable electricity generation and higher volume of renewable heat generation by 2020 would create a better investment climate for all low carbon technologies, including nuclear and CCS.'

There are clearly many possible mixes of energy options for electricity and heat supply that might prove to be viable, but arguably having a large element of inflexible plant will make it harder to balance the system, especially as the energy mix is likely to include an increasing amount of electricity from variable renewable sources. There is of course, conceivably, a possible alternative – a more or less completely nuclear-electric future, with nuclear electricity also being used for heating and to charge batteries for electric vehicles.

What next globally?

The debate on the way ahead in the UK still continues, but of course the UK is not alone, and as we have seen, increasingly nuclear vendors and operators like EDF are operating across national boundaries.

Although an attempt has been made to set the analysis in this book in a wider international context, the focus has mainly been on the prospects for a nuclear revival in the UK. The prognosis for the longer term globally is unclear. China, India and Japan have their own ongoing nuclear expansion programmes, and the USA, under President Bush, promoted a Global Nuclear Energy Partnership (see Chapter 15). Under this scheme, the US would help developing countries to build small modular plants, like the pebble-bed reactor, using sealed fuel capsules leased from the US. Once the fuel was used up, the capsules would be returned to the US for reprocessing, to extract the plutonium for use in a new fleet of US reactors. US Energy Secretary Samuel Bodman claimed that 'GNEP brings the promise of virtually limitless energy to emerging economies around the globe'. That may be rather risky overstatement, but, in theory, this approach would limit the risks of proliferation. However, it would mean that a lot of nuclear material would be in transit around the world, and the reprocessing activity would generate waste for the US to deal with. In general, around the world, reprocessing is falling out of favour. It has proved to be complex and expensive, and leads to more radiation exposure to workers and the public than any other part of the nuclear fuel cycle. Interestingly, in 2009, US president Obama cut funding for the work on reprocessing that was part of the USA's contribution to the GNEP programme.

Whether grand schemes like GNEP will prosper is therefore uncertain. It may be that, independently, for local strategic reasons, nuclear power will find an increasing role in some parts of the world outside of GNEP. But equally it may be seen as too expensive and risky.

At present there are mixed patterns of development. For example, China is certainly still pushing ahead with nuclear, as are India and South Korea and (evidently more contentiously) North Korea. Several Middle Eastern and north African countries have also indicated interests in going nuclear (perhaps most notably Iran). However in 2008 the South African company Eskom decided not to go ahead with a proposed nuclear plant, on the basis of the high costs. And in 2009, it was announced that the pebble-bed reactor project was to be halted due to financial problems. The nuclear expansion programme in Finland, based on the new European Pressurised-water Reactor (EPR), has met some difficulties with delays, and the expected construction cost has risen from €3 billion to €4.5 billion. The second EPR, being built in Flamanville in France, has also had significant cost overruns. As a result, the cost of power produced is likely to be around 20% more than planned, around 55 euros a megawatt hour, instead of the 46 euros/MWh announced when the project was launched in 2006.

In the USA, in his 'economic stimulus' funding in 2009, newly elected President Obama allocated around $160 billion to a range of renewable energy and efficiency projects and programmes, but did not provide the expected £50 billion in loan guarantees for new nuclear build. He also indicated that he would not support the planned waste repository at Yucca mountain. Moreover, as note above, he seems to have backed off from the idea of revisiting reprocessing as an option. While Obama has said that he is not opposed to nuclear if it can be made safer, so far his main focus seems to have been on other energy options.

While there are certainly signs that nuclear power is gaining more support, it clearly still remains a contentious issue worldwide. The nuclear lobby often points to France and more recently Finland, presumably as exemplars of what it would like to see globally. The UK may soon be added to that list. However, to the extent that options still remain open, there would seem to be a need for a continuing debate. Hopefully this book will help support that.

Index